SpringerBriefs in Computer Science

SpringerBriefs present concise summaries of cutting-edge research and practical applications across a wide spectrum of fields. Featuring compact volumes of 50 to 125 pages, the series covers a range of content from professional to academic.

Typical topics might include:

- A timely report of state-of-the art analytical techniques
- A bridge between new research results, as published in journal articles, and a contextual literature review
- A snapshot of a hot or emerging topic
- An in-depth case study or clinical example
- A presentation of core concepts that students must understand in order to make independent contributions

Briefs allow authors to present their ideas and readers to absorb them with minimal time investment. Briefs will be published as part of Springer's eBook collection, with millions of users worldwide. In addition, Briefs will be available for individual print and electronic purchase. Briefs are characterized by fast, global electronic dissemination, standard publishing contracts, easy-to-use manuscript preparation and formatting guidelines, and expedited production schedules. We aim for publication 8–12 weeks after acceptance. Both solicited and unsolicited manuscripts are considered for publication in this series.

More information about this series at http://www.springer.com/series/10028

Tayyaba Azim · Sarah Ahmed

Composing Fisher Kernels from Deep Neural Models

A Practitioner's Approach

 Springer

Tayyaba Azim
Center of Excellence in IT
Institute of Management Sciences
Peshawar, Pakistan

Sarah Ahmed
Institute of Management Sciences
Peshawar, Pakistan

ISSN 2191-5768 ISSN 2191-5776 (electronic)
SpringerBriefs in Computer Science
ISBN 978-3-319-98523-7 ISBN 978-3-319-98524-4 (eBook)
https://doi.org/10.1007/978-3-319-98524-4

Library of Congress Control Number: 2018950967

This Springer imprint is published by the registered company Springer Nature Switzerland AG
The registered company address is: Gewerbestrasse 11, 6330 Cham, Switzerland

To our creator ALLAH, the Almighty, beloved parents, dedicated mentors, loving friends and families…

Preface

This book aims at helping machine learning practitioners for successfully deploying deep models to learn better kernel functions, excessively used as a standard practice in the industry for decades. The recent surge of interest in deep learning has given rise to a plethora of literature demonstrating successful applications of deep models for various tasks in machine learning and computer vision. This move by the research community has compelled the industrial practitioners to update the existing technology behind their products and services to keep them ahead of the competition. However, the literature from the academia tends to focus more on novelty and proof of concept of deep learning algorithms on synthetic data sets rather than discussing the real-world challenges from an industrial practitioner's viewpoint. Keeping this gap in view, this book does not dive into the theoretical details of deep models and kernel methods; rather, it focuses on the essentials only which are necessary for the practitioners to learn for deploying advance deep learning algorithms in conjunction with the kernel methods. We have discussed the utility of deep models for enriching kernel-based learning and suggested the use of Fisher kernels to utilise the best of both the worlds.

Application of deep models for kernel methods brings computational and storage challenges for practitioners on real-world data sets and hence needs to be addressed explicitly. We have highlighted these challenges and suggested ways to solve them using off-the-shelf feature compression and selection approaches. In addition to providing guidance on how to connect the two competing paradigms, i.e. deep models and kernel methods, we have also summarised the necessary platforms, languages, toolboxes and data sets that can keep the practitioners going without entangling with the theoretical basis and proofs of deep models and kernel methods.

This book is written for the following target audience:

- Software engineers, developers and machine learning practitioners willing to use deep models or kernel methods in their products or services.
- Startups or companies seeking techniques to reduce the computational overhead of their technology.

- Graduate/undergraduate students of data mining, machine learning, data science, computer vision and statistical analysis courses can avail the book as a supplementary reading to quickly grasp and deploy the concepts of deep models and kernel methods.
- Curious readers who want to know the challenges of large-scale machine learning in the industry.

In order to render the task of compiling this book, we tried to incorporate all the groundbreaking fundamental and contemporary research on the topic, allowing our readers to understand the challenges as well as the scope of this genre of Fisher kernels. The suggestions given to the practitioners are based on research and theory tested through various programs to determine their effectiveness. Nevertheless, the readers are encouraged to download the cited toolboxes and data sets for getting hands-on experience of the problems and solutions in practice. We hope that the content covered satisfies all those with pragmatic engineering approach and adds to their confidence when designing intelligent solutions in the industry.

Many individuals and organisations have been a source of strength and direct/indirect inspiration that led to the compilation of this work. Among these include the ICPRAM 2017 organisers, in specific Marina Carvalho, NVIDIA for facilitating the hardware to support our experiments, Higher Education Commission (HEC) of Pakistan for providing start-up research grant that paved our way to this venture, the Institute of Management Sciences for encouraging faculty–student joint ventures and Springer's editorial team for guiding us through the entire process of publishing this work. A special note of gratitude goes to our parents and all mentors, in specific Prof. Mahesan Niranjan and Prof. Anwar Majid Mirza, for inculcating the spirit of determination, dedication, self-discipline, hard work, perseverance and lifelong learning in us.

Peshawar, Pakistan   Tayyaba Azim
July 2018   Sarah Ahmed

Acknowledgements

This book is the brainchild of a conversation between teacher and a research student reflecting upon the struggles of their younger selves in research and looking forward to help the next generations of young researchers and practitioners in machine learning to quickly learn from their experiences. It took us a great amount of time to collect, modify and simplify the important concepts shaping the field of machine learning and defining frontiers of success for science. This work is indebted to the invaluable contributions of many thoughtful, supportive and exceptionally skilled researchers and individuals who have attempted to benefit the community by proposing amazing algorithms, releasing open-source libraries and data sets and providing MOOCs to allow the community to learn new skills and develop their careers. We have utilised all these available resources to formulate the blueprint of this book and cited them to acknowledge their contribution. However, we truly believe that the authors of all the cited resources have taught us more than we could give them credit for over here.

We are extremely grateful to Allah, the Almighty, for giving us an opportunity and strength to pen down our understanding, thoughts and experiences on the topic. It would have been impossible to accomplish this goal without His benevolence and agreement. Next, we would like to express our gratitude to all those with whom we have had the pleasure to work with during this and related projects. In this regard, first author would specially like to thank her mentors Prof. Anwar Majid Mirza and Prof. Mahesan Niranjan for their valuable personal and professional guidance and teaching her a great deal about research and life in general.

We are much obliged to the Higher Education Commission of Pakistan and NVIDIA for their financial and hardware support that led to the commencement of this project. A special note of thanks goes to the ICPRAM 2017 organisers for giving us paper registration grant that led to the opportunities of sharing our research with the international research community as well as writing this SpringerBrief. We would also like to take this opportunity to thank our head of the institute, Dr. Muhammad Mohsin Khan, for encouraging educational joint ventures involving our youth, as well as all the technical and administrative staff at university, in specific, Dr. Iftikhar Amin and Syed Muhammad Nadir who helped us in resolving the power outage issues while executing this work. All our work colleagues, friends and critics who have inspired and motivated us directly or indirectly need not be forgotten here. Thank you all!

Our parents and families have remained extremely supportive and patient throughout the course of this journey. We would like to impart special thanks to our parents for their unconditional love, guidance, support and faith in our abilities to undertake this project. Their inspirational lives and presence as a role model have been a key driving factor in executing and completing this project. Finally, we would like to thank our publisher Springer, the associate editor Simon Rees and the project coordinator Manjula Saravanan for seizing the concept of this book with enthusiasm and guiding us through all the stages of writing and publishing this work. Any discrepancies, errors and shortfalls that may arise in the presented work could be a result of our own negligence. Kindly contact the authors directly at tayyaba.azim@imsciences.edu.pk and ssarahahmedd@gmail.com for notifying any errors, posing questions and providing feedback.

Contents

Acronyms

API	Application Programming Interface
CD	Contrastive Divergence
CPU	Central Processing Unit
DBM	Deep Boltzmann Machine
DBN	Deep Belief Network
FK	Fisher Kernel
FV	Fisher Vector(s)
GMM	Gaussian Mixture Model
GPU	Graphics Processing Unit
KL	Kullback–Leibler
MAP	Maximum a Posteriori
MCMC	Markov Chain Monte Carlo
MID	Mutual Information Difference
MIQ	Mutual Information Quotient
MKL	Multiple Kernel Learning
MRMR	Minimum Redundancy Maximum Relevance
MVG	Multivariate Gaussian Model
Parametric t-SNE	Parametric t-Distributed Stochastic Neighbour Embedding
RBM	Restricted Boltzmann Machine
SVMs	Support Vector Machines
VIF	Variance Inflation Factor

Chapter 1
Kernel Based Learning: A Pragmatic Approach in the Face of New Challenges

Abstract This chapter is not aimed at replacing literature on introduction to kernel methods or Fisher kernels. There are some excellent text books and tutorials on the topic by Schölkopf and Smola (Learning with kernels: support vector machines, regularization, optimization, and beyond. MIT Press (2002), [1]), Shawe-Taylor, Cristianini (Kernel methods for pattern analysis. Cambridge University Press (2004), [2]), Kung (Kernel methods and machine learning. Cambridge University Press, Princeton University (2014), [3])). In contrast to formal theory and proofs, this chapter briefly describes the evolution of kernel methods and the heuristics and methods that have helped kernel methods evolve over the past many years for solving the challenges faced by current machine learning practitioners and applied scientists.

Keywords Kernel functions · Kernel trick · Linear kernel · Non-linear kernels Mercer's theorem · Large scale machine learning

1.1 Kernel Learning Framework

The difficulty of expressing and solving complex real world problems through linear learning machines was first expressed by Marvin Minsky and Papert along with their allies in 1960s [4]. Detecting nonlinear relationship between data was a major point of concern for the researchers of the time who wanted to develop non-linear pattern recognition algorithms with the same level of computational efficiency and statistical guarantees as linear methods. This line of thought eventually led to a non-linear revolution in the mid 1980s when multiple layers of thresholded linear functions were introduced in multilayer neural networks to exploit more abstract representations of data. These non-linear algorithms provided a major breakthough in artificial intelligence, however they suffered from the problems of overfitting and local minima due to their reliance on greedy heuristics such as gradient descent learning algorithm. These problems were finally addressed by the researchers in mid 1990s with the introduction of kernel methods that analyse non-linear relationship of data with computational efficiency similar to that of linear methods. The proposed method satisfied the computational, statistical and conceptual requirements of the non-linear pattern analysis algorithms and overcome the issues of overfitting and local minima, typically associated with neural networks at that time.

Kernel methods provide a generic framework for analysing patterns of data by adopting a two step approach: The first step transforms the input feature space into another embedding space, whereas the second step consists of a learning algorithm that detects linear patterns in this new feature space. This strategy has shown fruitful results for various supervised and unsupervised learning tasks, purely because of two reasons: Detecting linear relationships is efficient computationally as well as statistically and has been explored thoroughly by the research community. Secondly, with the help of a trick defined via *kernel function*, it became easier to implicitly map data patterns in high dimensional spaces using *inner products* and detect linear relationship between them. Since its inception and the emergence of *kernel trick* in 1992, the kernel methods have benefited many existing and new learning algorithms such as support vector machines (SVMs), kernel fisher discriminant analysis (k-FDA), kernel principal component analysis (k-PCA), kernel regression, etc. Successful applications of kernel methods exist for a wide variety of problems spanning from object recognition, text categorization, natural language processing, time series prediction, gene expression analysis problems, etc.

1.1.1 Kernel Definition

We here elaborate the concept of kernel functions with a simple example. Formally, if we have data points $\mathbf{x}$ and $\mathbf{w} \in \mathbb{R}^n$, and a mapping function $\phi : \mathbb{R}^n \to \mathbb{F}$, we define the kernel function $K(\mathbf{x}, \mathbf{w})$ as follows:

$$K(\mathbf{x}, \mathbf{w}) = < \phi(\mathbf{x}), \phi(\mathbf{w}) >, \tag{1.1}$$

where $\phi(\mathbf{x})$ and $\phi(\mathbf{w})$ are the data point representations in some high dimensional feature space $\mathbb{F}$. The idea behind kernel functions is to embed original data $(\mathbf{x}, \mathbf{w})$ into high dimensional feature space where the linear relationship between data is detected at low computational cost.

Its important to note that all the variability and richness required to learn a powerful function class is introduced with the help of kernel functions. In essence, it is not the large dimensionality but the semantics of feature space that matter the most for pattern analysis and learning. This idea can be comprehended with the help of the following toy example demonstrated in Fig. 1.1. Data which is non-linearly separable in two dimensions gets linearly separable in higher dimensional feature space due to the function's discrimination power. We here see that learning in $\mathbb{F}$ is computationally simpler due to the application of linear decision boundary whereas this takes more time in the original data space $\mathbb{R}^n$ due to the utilisation of non-linear classifiers. Once a transformed representation is found with the help of kernel functions, a simple classification or regression algorithm could be applied to solve the problem at hand.

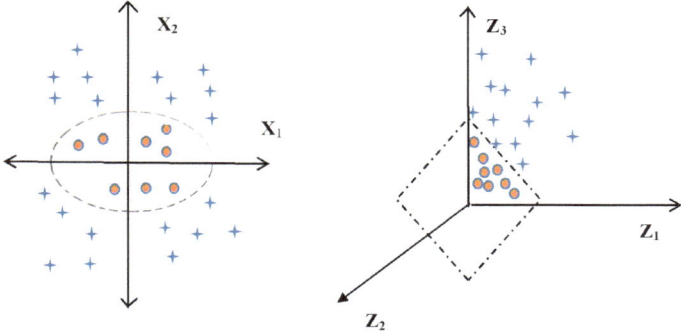

Fig. 1.1 In two dimensions, a complicated non-linear decision surface is required to separate the classes, whereas in the space of second order monomials, a linear classifier is sufficien:

1.2 Characteristics of Kernel Functions

In order for a function to be defined as a kernel, it must maintain the follcwing two important properties: (i) Symmetricity and (ii) Positive (semi-) definiteness. Kernels that satisfy *Mercer's theorem* are automatically said to comply to these requirements of kernel function. Mercer's theorem represents the inner product of data in some high dimensional embedded feature space sometimes referred to as Hilbert space H, though a large number of researchers require the additional properties of separability, completeness and infinite dimensions to approve this analogy. Kernels satisfy *positive semi-definiteness* property if all their eigen values are non-negative. The use of positive definiteness ensures that the optimisation problem is convex and therefore unique solution in terms of global minima exists.

It is important to note that many kernel functions which are not strictly positive definite have also shown to perform very well in practice. An example is the sigmoid kernel, which, despite its wide use, is not positive semi-definite for certain values of its parameters.

1.3 Kernel Trick

One of the most dominating reasons of the popularity of kernel functions is the use of *kernel trick*. If we want to transform our existing data into a higher dimensional feature space, which in many cases help us classify better (see Fig. 1.2), the kernel function enables one to implicitly operate in a higher dimensional feature space without ever computing the coordinates of the data in that space. With the help of *kernel trick*, instead of computing the exact transformation of our data; just the inner product of the data in that higher dimensional space is computed. It is a lot easier to get the inner product in a higher dimensional space than the actual points in a higher dimensional space. The kernel trick enables one to substitute any dot product with a kernel function. Thus, wherever a dot product is found in the formulation of a machine learning algorithm, it could be replaced with a kernel function. Mathematically,

$$\text{Kernel Trick } :< x_i, x_j >= \sum_{i,j=1}^{d} x_i . x_j \qquad (1.2)$$

can be replaced with $K(x_i, x_j)$ for any valid kernel function K.   (1.3)

It is due to this property of the kernel methods, many linear supervised and unsupervised learning algorithms have been kernelized including ridge regression, perceptrons, Fisher discriminant analysis, principal component analysis (PCA), k-means clustering, and independent component analysis (ICA).

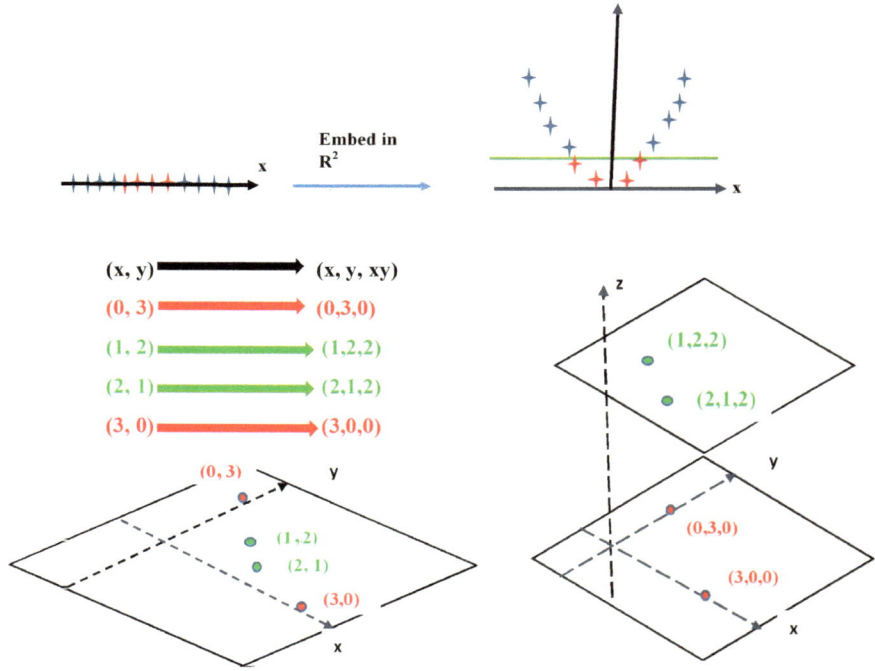

Fig. 1.2 Demonstrating binary classification problem of two different sets of points which are not linearly separable in the original feature space of two dimensions. The data becomes *linearly separable* in the new representation space mapped through kernel functions

1.4 Types of Kernel Functions

Following are some of the most popular kernel functions used for solving different real world problems:

- Linear Kernels
- Gaussian Kernels
- Polynomial Kernels
- Sigmoid Kernels
- Hyperbolic Tangent Kernels
- String Kernels
- Fisher Kernels
- Graph Kernels

Kernels can also be classified as anisotropic stationary, isotropic stationary, compactly supported, locally stationary, non-stationary or separable non-stationary. Another means of categorising kernels is by labelling them as scale-invariant or scale-dependant, which is an interesting property as scale-invariant kernels drive the training process invariant to scaling of the data.

The selection of an appropriate kernel and its parameters is often based on experience or a potentially costly search. Empirical studies have shown that the choice of kernel often affects the resulting performance of kernel methods significantly. In fact, inappropriate kernels can result in sub-optimal or very poor performance. For many real-world situations, it is often not an easy task to choose an appropriate kernel function, which usually may require some domain knowledge that would be difficult for non-expert users. To address such limitation, recent years have witnessed the active research of learning effective kernels automatically from data [5]. One popular example technique for kernel learning is multiple kernel learning (MKL) [5, 6], which aims at learning a linear (or convex) combination of a set of predefined kernels in order to identify a good target kernel for the applications. Comparing with traditional kernel methods using a single fixed kernel, MKL does exhibit its strength of automated kernel parameter tuning and capability of concatenating heterogeneous data. Over the past few years, MKL has been actively investigated, in which a number of algorithms have been proposed to resolve the efficiency of MKL [5, 7–9], and a number of extended MKL techniques have been proposed to improve the regular linear MKL method [5, 10–12]. Some researchers have tried to extend kernels to deep learning as well [13, 14].

1.5 Challenges Faced by Kernel Methods and Recent Advances in Large-Scale Kernel Methods

For any given data set with n elements, the kernel function, K is a $n \times n$ symmetric matrix, i.e.

$$K(\mathbf{x}, \mathbf{y}) = K(\phi(\mathbf{x}), \phi(\mathbf{y})). \tag{1.4}$$

Although computation of n^2 elements of the kernel seem an easy quadratic problem, an in depth analysis reveals that it is quite expensive in terms of the required computational memory. As the data set scales up, the matrix grows exponentially large to fit inside the available memory. In this case, the kernel coefficients are either computed on the fly or are stored in the cache memory. The kernel cache hit rate contributes to the overall training time of the kernel and is therefore considered as an expensive operation for large data sets.

The computational complexity of the kernel matrix has often been a barrier in scaling kernel methods up for large scale learning problems. There are a few successful applications of kernel methods at large scale comparable to speech and image recognition problems. It is due to this computational cost, many people believe that it may be difficult, if not impossible for the kernel methods to compete deep learning methods on large scale learning problems. Despite this prevailing sentiment, there has been a recent surge of interest in the kernel research community [15] to empower the kernel methods for large scale learning problems. Some of the recent advances that have accelerated the kernel methods in comparison to deep learning paradigm are discussed ahead. Instead of computing the entire kernel matrix, its approximations are created with the help of low-rank matrices through methods such as incomplete Cholesky decomposition, Lanczos approximation, Nystrom method, the fast Gaussian transform and FALCON [16–19]. For classification problems, the optimisation problem is often solved by using cutting plane [20] or dual coordinate decent technique [21].

References

1. Schölkopf, B., Smola, A.: Learning with Kernels: Support Vector Machines, Regularization, Optimization, and Beyond. MIT Press, USA (2002)
2. Shawe-Taylor, J., Cristianini, N.: Kernel Methods for Pattern Analysis. Cambridge University Press, New York (2004)
3. Kung, S.: Kernel Methods and Machine Learning. Cambridge University Press, Princeton University, New Jersey, USA (2014)
4. Mirsky, M., Papert, S.: Perceptrons: An Introduction to Computational Geometry. MIT Press, USA (2017)
5. Bach, F., Lanckriet, G., Jordan, M.: Multiple kernel learning, conic duality, and the SMO algorithm. In: Proceedings of the Twenty-First International Conference on Machine Learning (2004)
6. Sonnenburg, S., Rätsch, G., Schäfer, C.: A General and efficient multiple kernel learning algorithm. In: Advances in Neural Information Processing Systems, pp. 1273–1280 (2006)
7. Sonnenburg, S., Rätsch, G., Schäfer, C., et al.: Large scale multiple kernel learning. J. Mach. Learn. Res. **7**, 1531–1565 (2006)
8. Xu, Z., Jin, R., King, I., Lyu, M.: An extended level method for efficient multiple kernel learning. In: Advances in Neural Information Processing Systems, pp. 1825–1832 (2009)
9. Rakotomamonjy, A., Bach, F., Canu, S., et al.: More efficiency in multiple kernel learning. In: Proceedings of the 24th International Conference on Machine Learning, pp. 775–782. ACM (2007)
10. Gehler, P., Nowozin, S.: Infinite kernel learning. Technical report, TR-178 (2008)

11. Gönen, M., Alpaydin, E.: Localized multiple kernel learning. In: Proceedings of the 25th International Conference on Machine Learning, pp. 352–359. ACM (2008)
12. Varma, M., Babu, R.: More generality in efficient multiple kernel learning. In: Proceedings of the 26th Annual International Conference on Machine Learning, pp. 1065–1072. ACM (2009)
13. Cho, Y., Saul, L.: Kernel methods for deep learning. In: Advances in Neural Information Processing Systems, pp. 342–350 (2009)
14. Zhuang, J., Tsang, I., et al.: Two-layer multiple kernel learning. In: Proceedings of the Fourteenth International Conference on Artificial Intelligence and Statistics, pp. 909–917 (2011)
15. Lu, Z., May, A., Liu, K., et al.: How to scale up kernel methods to be as good as deep neural nets (2014). arXiv preprint arXiv:1411.4000
16. Wu, J., Zheng, W., Lai, J.: Approximate kernel competitive learning. Neural Netw. 63, 117–132 (2015)
17. Pourkamali, F., Becker, S.: Randomized clustered nystrom for large-scale kernel machines (2016). arXiv preprint arXiv:1612.06470
18. Rahimi, A., Recht, B.: Random features for large-scale kernel machines. In: Advances in Neural Information Processing Systems, pp. 1177–1184 (2008)
19. Rudi, A., Carratino, L., Rosasco, L.: FALKON: An optimal large scale kernel method. In: Advances in Neural Information Processing Systems, pp. 3891–3901 (2017)
20. Franc, V., Sonnenburg, S.: Optimized cutting plane algorithm for large-scale risk minimization. J. Mach. Learn. Res. 2157–2192 (2009)
21. Hsieh, C., Chang, K., Lin, C., et al.: A dual coordinate descent method for large-scale linear SVM. In: Proceedings of the 25th International Conference on Machine Learning, pp. 408–415. ACM (2008)

Chapter 2
Fundamentals of Fisher Kernels

Abstract This chapter introduces a specific genre of kernels that draws a formal connection between the *generative* and *discriminative models* of learning. Both the paradigms offer unique complementary advantages over one another, yet there always existed a need to combine the best of both the worlds for solving complex problems. This gap was filled by Tommy Jaakola through the introduction of *Fisher* kernels in 1998 and since then it has played a key role in solving problems from computational biology, computer vision and machine learning. We introduce this concept here and show how to compute Fisher vector encodings from deep models using a toy example in MATLAB.

Keywords Fisher kernel · Generative models · Discriminative models
Multivariate Gaussian (MVG) model · Gaussian mixture model (GMM)
Fisher vectors · Fisher vector normalisation · k-Nearest neighbour
Fisher information matrix · L2 normalisation

2.1 Introduction

There is a fundamental difference in the doctrine of generative and discriminative learning models making them relevant and useful in different situations. *Generative models* are most importantly known for their data *modelling power* and offer the advantage of dealing with data of variable length, such as speech, vision, text and biomedical sequences which are often vectors of variable size and face difficulty in a simple classification framework. After training, the generative models can further lend themselves to Bayesian rule for classification task. The generative models are also scalable, yet their major deficiency lies in their inability to discriminate between data samples. In contrast, the *discriminative models* are trained with the primary objective of improving the classification performance, however the framework requires fixed length data sequences and optimal choice of *kernel function* and its parameters to reveal its discriminatory strength. The process of kernel selection and its parameters is a matter of experience or a potentially expensive search that leads us to satisfactory classification results. Empirical studies have shown the supremacy

© The Author(s), under exclusive licence to Springer Nature Switzerland AG 2018
T. Azim and S. Ahmed, *Composing Fisher Kernels from Deep Neural Models*,
SpringerBriefs in Computer Science, https://doi.org/10.1007/978-3-319-98524-4_2

of discriminative classifiers over the generative models, yet the contrasting features of both the paradigms make one yearn for a solution that combines the benefits of both the approaches. This gap between the two approaches is bridged by a hybrid generative-discriminative method of classification that deploys *Fisher kernels* for the task.

2.2 The Fisher Kernel

Fisher kernels provide a systematic way of using the parameters of the generative model to define an embedding space for kernels capable of being deployed in discriminative classifiers. The Fisher kernel finds out the similarity between any two examples, $\mathbf{x}_i$ and $\mathbf{x}_j$ as follows:

$$K(\mathbf{x}_i, \mathbf{x}_j) = U_{\mathbf{x}_i}^T \mathbf{F}^{-1} U_{\mathbf{x}_j}, \tag{2.1}$$

where the Fisher score, $U_{\mathbf{x}_i}$ maps the data $\mathbf{x}_i$ into a point in the gradient space of the manifold M_θ, also regarded as a *Fisher space*. The Fisher score is mathematically described as:

$$\mathbf{U}_{\mathbf{x}} = \nabla_\theta \log P(\mathbf{x}|\boldsymbol{\theta}), \tag{2.2}$$

where $\boldsymbol{\theta}$ refers to the set or vector of generative model's parameters and $\log P(\mathbf{x}|\boldsymbol{\theta})$ defines the log-likelihood function of the data learnt by the generative model. The gradient of the log-likelihood function with respect to each model parameter describes how that parameter contributes to the process of generating the observed sample $\mathbf{x}$ and hence captures the generative process of a sequence much better than just the posterior probabilities. If the gradient vectors are similar, it means the two data points would adapt the model in the same way and vice versa. Thus the separability of different data points is accomplished by taking into account the data generation process through their underlying probability distributions. The *Fisher information matrix* $\mathbf{F}$ in Eq. 2.1 tells us about the covariance of the Fisher scores defined by $U_{\mathbf{x}}$ as:

$$\mathbf{F} = \mathrm{E}_{P(\mathbf{x}|\theta)}[U_{\mathbf{x}} U_{\mathbf{x}}^T] \tag{2.3}$$

For computational simplicity, this covariance matrix, $\mathbf{F}$ is often ignored in practice by taking it as an identity matrix $\mathbf{I}$, leading to non-invariant Fisher kernel shown below:

$$K(\mathbf{x}, \mathbf{x}^T) = U_{\mathbf{x}_i}^T U_{\mathbf{x}_j} \tag{2.4}$$

Empirical results reveal that the use of the invertible information matrix is immaterial [1], however some researchers have either used a diagonal matrix [2, 3] or an approximation of the Fisher kernel [4] to get better discrimination results (Fig. 2.1).

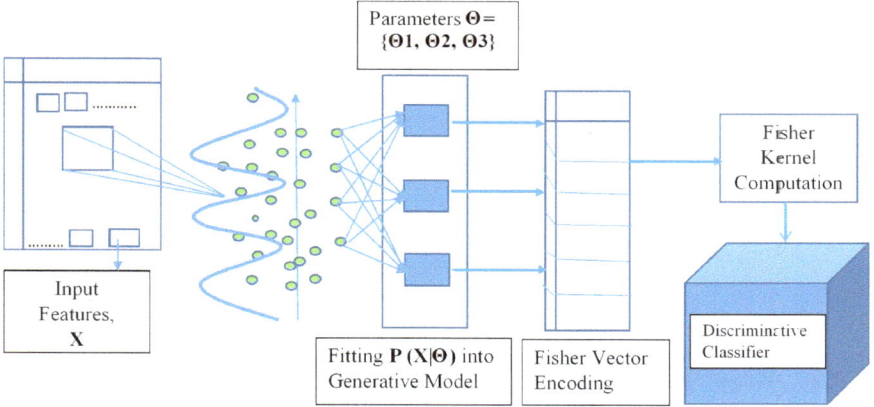

Fig. 2.1 Fisher kernel framework combining the benefits of generative and discriminative models. For image classification, the features from images are extracted and modelled with the help of a generative probability distribution that leads to Fisher scores and thus Fisher kernel ultimately. The Fisher kernel could be embedded into any discriminative classifier like support vector machines (SVM) and linear discriminant analysis (LDA)

2.2.1 Fisher Vector Normalisation

Since the Fisher vector encodings are dominantly used in large margin classifiers like support vector machines (SVM), it is highly recommended to normalise the features because SVM is highly sensitive to the way features are scaled. The goal of normalisation is to bring the values of features into the same scale without losing any information and to model the data correctly. If the data is un-normalised then due to the great difference among the scale of features, the learning algorithm can show a knock-on-effect on performance. The Fisher encodings have been normalised using different techniques such as L1/L2 normalisation, power normalisation and min-max normalisation [5, 6] to yield better classification performance. Another method to scale the features is to normalise the Fisher kernel function after constructing it from the unnormalised Fisher scores. The use of Fisher information matrix in kernel formulation is also a means to normalise the Fisher scores for improving the quality of extracted Fisher encodings for the classification task.

2.2.1.1   Matlab Implementation of Min-Max Normalisation

Min-Max normalisation technique normalises the features in the range [0, 1]. If **x** is an n-dimensional feature vector, the Min-Max normalisation is computed by using the following linear interpretation formula:

$$\mathbf{x}_{norm} = (x_i - x_{min})/(x_{max} - x_{min}), \tag{2.5}$$

where x_{min} and x_{max} represent the minimum and maximum values across all dimensions for each data vector, $\mathbf{x}$ respectively.

```
%Min-Max Normalisation
train_min=min(train_data);  %calculate min of training/test data
train_max=max(train_data);  %calculate max of training/test data
t=train_max-train_min;%subtract min(data)from max(data)
%loop to subtract each element of data with min(x)and divide by max(x)-min(x)
for i=1:size(train_data,1)
  normalized_traindata(:,i)=(train_data(:,i)-train_min(i))/t(i);
end
```

2.2.1.2 Matlab Implementation of L2-Normalisation

The Euclidean or L2-norm of an n-dimensional column vector $\mathbf{x}$ is defined as follows:

$$\| \mathbf{x} \|_2 = \sqrt{\sum_{k=1}^{n} x_k^2} \tag{2.6}$$

```
%Perform L2 Normalization
v= (sum(train_data.^2)).^(1 /2);%compute l2 norm of each column
normalized_traindata = bsxfun (@times,train_data,1 ./ v);
```

2.2.2 Properties of Fisher Kernels

There are a few properties of Fisher kernel function which are stated in the form of the following theorem [1]:

Theorem 1 *For any (suitably regular [1]) probability model $P(\mathbf{x}|\boldsymbol{\theta})$ with parameters $\boldsymbol{\theta}$, the Fisher kernel: $K(\mathbf{x}_i, \mathbf{x}_j) = U_{\mathbf{x}_i}^T F^{-1} U_{\mathbf{x}_j}$, where $U_{\mathbf{x}} = \nabla_\theta \log P(\mathbf{x}|\boldsymbol{\theta})$ has the following properties:*

- *It is a valid kernel function.*
- *It is invariant to any invertible (and differentiable) transformation of the parameters.*
- *A kernel classifier employing the Fisher kernel derived from a model that contains the label as a latent variable is, asymptotically, at least as good a classifier as the maximum-a-posteriori (MAP) labelling based on the model.*

The first property is immediate since the Fisher information matrix F is symmetric as well as positive definite satisfying the requirements of an inner product space for the Fisher kernel. The second property follows from the fact that the kernel is defined on the manifold M_θ, that uses as a feature space the gradients of the log likelihood of the probability distribution with respect to its parameters rather than the model parameters itself. The third property can be established on the basis of the discriminative derivation[1] of this kernel shown in [1].

2.2.3 Applications of Fisher Kernels

Fisher kernels were first derived from hidden Markov models (HMM) to show significant improvement over previous methods for protein domain classification [1, 7]. The idea got an immediate attention of the research community who utilised it for different applications in biology, speech, vision, web and text to improve state of the art classification and prediction performances. Some of the other generative models deployed for Fisher kernel extraction are: Multivariate Gaussian model [8], Gaussian mixture models [9, 10], probabilistic hierarchical clustering model [11], probabilistic latent semantic indexing [12], latent Dirichlet allocation model [13], Dirichlet compound multinomial (DCM) distribution [14], restricted Boltzman machine [8], deep Boltzmann machine [15] factored RBM [16] and generative convolution neural networks [17].

2.2.4 Illustration of Fisher Kernel Extraction from Multivariate Gaussian Model

To understand the concept of Fisher kernels, lets consider the example of drawing Fisher kernel from multivariate Gaussian model, $N(\mathbf{x}, \boldsymbol{\Sigma})$, whose probability distribution is given as:

$$P(\mathbf{x}|\boldsymbol{\mu}, \boldsymbol{\Sigma}) = \left(\frac{1}{(2\pi)^{d/2}|\boldsymbol{\Sigma}|^{1/2}}\right) \exp\left(\frac{-1}{2}(\mathbf{x} - \boldsymbol{\mu})^T (\boldsymbol{\Sigma})^{-1} (\mathbf{x} - \boldsymbol{\mu})\right) \quad (2.7)$$

Assuming that the data samples $X = \{\mathbf{x_1}, \mathbf{x_2}, \ldots, \mathbf{x_N}\}$ are independent and identically distributed, we can form the joint distribution of the N samples via the following likelihood function:

---

[1] We must have twice differentiable likelihood so that the Fisher information I exists and I must be positive definite at the chosen θ.

$$\log \mathscr{L} = \log \prod_{i=1}^{N} (P(\mathbf{x}_{(i)} | \boldsymbol{\mu}, \boldsymbol{\Sigma}))$$

$$\log \mathscr{L} = \log \prod_{i=1}^{N} \left[\left(\frac{1}{(2\pi)^{d/2} |\boldsymbol{\Sigma}|^{1/2}} \right) \exp \left(\frac{-1}{2} (\mathbf{x}_i - \boldsymbol{\mu})^T (\boldsymbol{\Sigma})^{-1} (\mathbf{x}_i - \boldsymbol{\mu}) \right) \right],$$

$$\log \mathscr{L} = - \frac{Nd}{2} \log(2\pi) - \frac{N}{2} \log(|\boldsymbol{\Sigma}|) - \frac{\sum_{i=1}^{N} (\mathbf{x}_i - \boldsymbol{\mu}) \boldsymbol{\Sigma}^{-1} (\mathbf{x}_i - \boldsymbol{\mu})}{2},$$

where $\boldsymbol{\mu}$ is a d-dimensional mean vector, $\boldsymbol{\Sigma}$ is a $d \times d$ dimensional covariance matrix, and $|\boldsymbol{\Sigma}|$ denotes the determinant of the covariance matrix $\boldsymbol{\Sigma}$. Since the parameters of this model are $\boldsymbol{\theta} = \{\boldsymbol{\mu}, \boldsymbol{\Sigma}\}$, the Fisher score $U_{\mathbf{x}}$ for the model is calculated as below:

$$U_{\mathbf{x}} = \begin{bmatrix} \nabla_{\boldsymbol{\mu}} \log \mathscr{L} \\ \nabla_{\boldsymbol{\Sigma}} \log \mathscr{L} \end{bmatrix} \text{ where} \tag{2.8}$$

$$\nabla_{\boldsymbol{\mu}} \log \mathscr{L} = \boldsymbol{\Sigma}^{-1} (\mathbf{x} - \boldsymbol{\mu}) \tag{2.9}$$

$$\nabla_{\boldsymbol{\Sigma}} \log \mathscr{L} = \frac{1}{2} \left[-\boldsymbol{\Sigma}^{-1} + \boldsymbol{\Sigma}^{-1} (\mathbf{x} - \boldsymbol{\mu})(\mathbf{x} - \boldsymbol{\mu})^T \boldsymbol{\Sigma}^{-1} \right]. \tag{2.10}$$

2.2.5 Illustration of Fisher Kernel Derived from Gaussian Mixture Model (GMM)

For the sake of simplicity of demonstration, we assume a two Gaussian mixture model whose probability distribution for all n data points is given as below. This formulation could be extended to a mixture of any number of Gaussian generative models.

$$P(\mathbf{x}_n | \boldsymbol{\theta}) = w_1 N_{[n,1]}(\mu_1; \Sigma_1) + w_2 N_{[n,2]}(\mu_2; \Sigma_2), \text{ where} \tag{2.11}$$

$$N_{[n,k]}(\boldsymbol{\mu}_k, \Sigma_k) = \frac{(2\pi)^{\frac{-N}{2}} \exp^{\frac{-1}{2}(\mathbf{x}_n - \mu_k)^T \Sigma_k^{-1}(\mathbf{x}_n - \mu_k)}}{|\Sigma_k|^{\frac{1}{2}}} \tag{2.12}$$

The log-likelihood function of the data for GMM is given as below:

$$\log \mathscr{L} = \log \prod_{i=1}^{n} (P(\mathbf{x}_i | \boldsymbol{\mu}, \boldsymbol{\Sigma})) \tag{2.13}$$

This implies that the gradient of the log likelihood of Gaussian mixture model with respect to its parameters $\boldsymbol{\theta} = \{\boldsymbol{\mu_1}, \boldsymbol{\mu_2}, \Sigma_1, \Sigma_2, w_1, w_2\}$ is given as:

$$\nabla_\theta \log \mathcal{L} = \begin{bmatrix} \nabla_{\mu_1} \log \mathcal{L} \\ \nabla_{\mu_2} \log \mathcal{L} \\ \nabla_{\Sigma_1} \log \mathcal{L} \\ \nabla_{\Sigma_2} \log \mathcal{L} \\ \nabla_{w_1} \log \mathcal{L} \\ \nabla_{w_2} \log \mathcal{L} \end{bmatrix}, \tag{2.14}$$

$$\nabla_\theta \log \mathcal{L} = \begin{bmatrix} \gamma_{[n,1]} S_{[n,1]} \\ \gamma_{[n,2]} S_{[n,2]} \\ \gamma_{[n,1]} Q_{[n,1]} \\ \gamma_{[n,2]} Q_{[n,2]} \\ \frac{\gamma_{[n,1]}}{w_1} \\ \frac{\gamma_{[n,2]}}{w_2} \end{bmatrix}, \tag{2.15}$$

$$\text{where } \gamma_{[n,k]} = \frac{w_k N_{[n,k]}(\mu_k, \Sigma_k)}{w_1 N_{[n,1]}(\mu_1; \Sigma_1) + w_2 N_{[n,2]}(\mu_2; \Sigma_2)}, \tag{2.16}$$

$$S_{[n,k]} = (x_n - \mu_k)^T \Sigma_k^{-1}, \tag{2.17}$$

$$Q_{[n,k]} = \frac{1}{2} \left[-\left[\text{vec}(\Sigma_k^{-1}) \right]^T + S_{[n,k]} \otimes S_{[n,k]} \right], \tag{2.18}$$

$$\text{vec}(F) = [f_{11}, f_{12}, \ldots, f_{mn}]^T . \tag{2.19}$$

2.2.5.1 MATLAB Implementation of Fisher Kernel Derived From Gaussian Mixture Model

Gaussian mixture model (GMM) has been widely used to draw Fisher kernel for improving the performance of large scale visual classification and retrieval tasks. The model could be trained on any set of features, however the most successful features used to train the probabilistic model are local descriptors, i.e. scale invariant feature transform (SIFT). To demonstrate the learning of GMM, we have randomly generated a data set of 1000 samples in 2 dimensions but one can instead use SIFT or any other features for model training too.

```
%Create Data-set
number_samples = 1000 ;
number_dimensions = 2 ;
data = rand(number_samples,number_dimensions) ;
labels = randi(2, 1000,1);
```

Fit Gaussian Mixture Model on Data Set

In order to train a Gaussian mixture model for learning optimal parameters on the data set, expectation maximisation (EM) algorithm is used. The function returns the trained model parameters θ. In the code snippet shown below, mean of all the Gaussian distributions, weights assigned to different Gaussian distributions and diagonal of corresponding covariance matrices is returned.

```
%Fit GMM on Data-set
K=16; %number of Gaussians
[weights,mean,sigma]=compute_gmm(data,k);
```

Compute Fisher Kernel from GMM

Once all the parameters of GMM model are computed through training, Fisher vector encodings (Eq 2.5) could be derived by calling the function gmm_Fisher().

```
%Compute Fisher Vector Encoding
[Fisher_vectors] = gmm_Fisher(data,weights,mean,sigma);
```

Fisher Vector Normalisation

The Fisher vector encodings are next normalised using any of the normalisation options that can substantially affect the performance of the classifiers or predictors.

```
%Fisher Vector Normalization
Option='min-max';% You can change the option value by L1-norm,L2-norm
[normalized_Fishervector]=Fisher_normalize(Fisher_vectors,option);
```

Computing Fisher Kernel and Using a Standard Classifier

After the normalisation of Fisher vectors, the Fisher vectors could be embedded into any classifier deploying kernels (like support vector machines) or passed on to standard classifier (like k-nearest neighbour) for performance estimation.

```
%Split Dataset into Train and Test Set
[train_data,test_data,train_labels,test_labels]=split(normalized_Fishervector,labels);
%Perform 1-Nearest Neighbor Classifier
c=knnclassify(test_data,train_data,train_labels);
cp=classperf(c,test_labels);
cp.CorrectRate;%Accuracy rate
```

References

1. Jaakkola, T., Haussler, D.: Exploiting generative models in discriminative classifiers. In: Advances in Neural Information Processing Systems, pp. 487–493 (1999)
2. Nyffenegger, M., Chappelier, J., Gaussier, É.: Revisiting fisher kernels for document similarities. In: European Conference on Machine Learning, pp. 727–734, Springer (2006)

3. Perronnin, F., Dance, C.: Fisher kernels on visual vocabularies for image categorization. In: IEEE Conference on Computer Vision and Pattern Recognition, pp. 1–8, IEEE (2007)
4. Maaten, L.: Learning discriminative fisher kernels. In: ICML **11**, 217–224 (2011)
5. Perronnin, F., Sánchez, J., Mensink, T.: Improving the fisher kernel for large scale image classification. In: European Conference on Computer Vision, pp. 143–156, Springer (2010)
6. Ahmed, S., Azim, T.: Compression techniques for deep fisher vectors. In: ICPRAM, pp. 217–224 (2017)
7. Jaakkola, T., Diekhans, M., Haussler, D.: A discriminative framework for detecting remote protein homologies. J. Comput. Biol. **7**(1–2), 95–114 (2000)
8. Azim, T., Niranjan, M.: Inducing discrimination in biologically inspired models of visual scene recognition. In: IEEE International Workshop on Machine Learning for Signal Processing, (MLSP) (2013)
9. Smith, N., Niranjan, M.: Data-dependent kernels in SVM classification of speech patterns. Technical Report, Cambridge University (2001)
10. Moreno, P., Rifkin, R.: Using the fisher kernel method for web audio classification. In: Acoustics, Speech, and Signal Processing (ICASSP), vol. 4, pp. 2417–2420 (2000)
11. Vinokourov, A., Girolami, M.: Document Classification Employing the Fisher Kernel Derived from Probabilistic Hierarchic Corpus Representations. Springer (2001)
12. Chappelier, J., Eckard, E.: PLSI: The true fisher kernel and beyond. European Conference on Machine Learning and Knowledge Discovery in Databases: Part **I**, 195–210 (2009)
13. Sun, Q., Li, R., Luo, D., et al.: Text segmentation with LDA based fisher kernel. In: Proceedings of the 46th Annual Meeting of the Association for Computational Linguistics on Human Language Technologies: Short Papers, Association for Computational Linguistics, pp. 269–272 (2008)
14. Elkan, C.: Deriving TF-IDF as a fisher kernel. In: 12th International Conference on String Processing and Information Retrieval (SPIRE), pp. 296–301 (2005)
15. Azim, T.: Fisher kernels match deep models. Electron. Lett. **53**(6), 397–399 (2017)
16. Azim, T., Niranjan, M.: Texture classification with Fisher kernel extracted from the continuous models of RBM. In: International Conference on Computer Vision Theory and Applications (VISAPP), vol. 2, pp. 684–690, IEEE (2014)
17. Songa, Y., Hong, X., McLoughlin, I.: Image classification with CNN based fisher vector coding. In: Visual Communications and Image Processing (VCIP), pp. 1–4 (2016)

Chapter 3
Training Deep Models and Deriving Fisher Kernels: A Step Wise Approach

Abstract Deep networks and Fisher kernels are two competitive approaches showing strides of progress and improvement for computer vision tasks in specific the large scale object categorisation problem. One of the recent developments in this regard has been the use of a hybrid approach that encodes higher order statistics of deep models for Fisher vector encodings. In this chapter we shall discuss how to train a deep model for extracting Fisher kernel. The tips discussed here are validated by industrial practices and research community through mathematical proofs by LeCun et al. (Neural networks: tricks of the trade. Springer, pp 9–50 (1998), [1]), Bengio (Neural networks: tricks of the trade. Springer, pp 437–478 (2012), [2]) and case studies.

Keywords Deep models · Restricted Boltzmann machine (RBM)
Deep Boltzmann machine (DBM) · Hyper-parameter tuning · Regularisation
Fisher scores · Support vector machines (SVM) · Min-max normalisation

3.1 How to Train Deep Models?

In order to derive a Fisher kernel, one needs to have a generative model first so that Fisher based encodings could be extracted. Training deep models requires certain amount of practice for tuning hyper-parameters such as learning rate, mini-batch size, selection of appropriate activation function, number of hidden layers, number of hidden units in each layer, weight initialisation values. In the sections below, we shall provide some guidance to practitioners for training deep models with good hyper-parameters and show how to make them deployable by discriminative classifiers ahead.

3.1.1 Data Preprocessing

Many machine learning practitioners are persistent on throwing raw features to any deep model. It presumably gives good performance by throwing raw features in a simple architecture but for complex models it may or may not work. So, it is important that when you are working with natural language processing, tackling computer vision problems or doing any statistical modeling etc., always try to preprocess your data. Not only can this reduce your computational overhead and memory requirements but shall guarantee you consistent outputs. There are several data pre-processing steps one may take dependent upon the problem and data at hand:

- Remove any training example that holds corrupted data. Here corruption refers to the data that is incomplete, may hold an incorrect label or do not possess the data you require in addressing the problem at hand.
- If the data set contains limited training examples, state of the art deep models with millions of parameters are more likely to overfit. Perform *data augmentation* to increase the amount of the training data. Conventional techniques of data augmentation include the application of affine transformations to the original data set image by translating, flipping, color jittering, changing the color palette of image. Recently researchers have also used neural networks such as generative adversarial networks (GAN) to augment the data [3, 4].
- Perform *data normalisation* so that all the attributes/features appear in the same range. There are different normalisation techniques such as power normalisation, z-score normalisation, L1/L2 normalisation, min-max normalisation. In deep architectures, neurons are excited by positive weights and inhibited by the negative weights. In such models, min-max normalisation scheme that re-scales the data in positive range [0, 1] has shown good classification results [5].

3.1.2 Selection of an Activation Function

One of the essential components of any deep architecture is the type of *activation function* deployed in each neuron. The main purpose of an activation function is to introduce non-linearity in the deep network by converting each incoming input into a non-linear output that may/may not fire a neuron. The result is dependent upon the type of mathematical operation performed on the input features. If we do not have an activation function, the bias and weights of a network shall simply perform linear transformation of the input which has limited capacity to solve complex problems. It is only because of the non-linear activation function that neural networks are regarded as universal function approximators. Some of the most popular activation functions are sigmoid, hyperbolic tangent [6], rectified linear unit(ReLU), leaky Relu, etc., where some have shown to be more successful [7–9] than the rest.

Choosing the right activation function depends upon the nature of the problem we intend to solve. There is no rule of thumb for this selection, yet there are a few guidelines gathered from the experience of industry and academia:

- The derivative of linear activation function is constant, so the gradient would be the same when we do back propogation. Using such activation functions is only ideal when performing simple tasks or where interpretability is highly required.

- Sigmoid function is a non-linear activation function that takes a real number as input and squashes the output in the range [0, 1]. This is one of the oldest activation functions used in neural networks but has two limitations: One, its outputs are not zero-centered and second, it kills and saturates the gradients due to which neurons and its associated weights are not updated. It is due to this saturating gradient issue that sigmoid functions are not preferred in ultra deep models.
- The hyperbolic tangent activation function is also non-linear and provides zero-centered output in the range [−1, 1]. The gradients of hyperbolic tangent function are stronger, hence its easier to optimize the model parameters using this activation function. This function is preferred over sigmoid function, however it also has the limitation of vanishing gradient resulting in saturated neurons.
- Rectified linear unit (ReLU) is the most popular and commonly used activation function in neural architectures because it does not suffer from the problem of vanishing gradient. This non-linear function gives an output only when its input is positive, otherwise the neurons do not fire. ReLU is non-differentiable at 0 which in-turn creates dead neurons because zero gradients flow through them during back propagation making them inactive for all inputs.
- Leaky ReLU or Max out is the improved version of the ReLU function and gives a solution to dead neuron problem. LeakyReLu enforces a small negative gradient flow through the network when the unit is not active. Parametric ReLu takes this idea further by learning this negative coefficient as a parameter that is learnt along with the parameters of the neural network.

Given the pros and cons of the above mentioned activation functions, ReLu appears to be the best choice to take a start with. The activation function gives sparse representations due to fewer neurons firing. As a rule of thumb, it should be used in hidden layers and after the convolution layer. If the network suffers from dead neuron problem, one can replace ReLU with Maxout/ Leaky ReLU.

3.1.3 Selecting the Number of Hidden Layers and Hidden Units

Determining the number of hidden layers and selecting the number of hidden units in each hidden layer is one of the most important decisions that determine our neural architecture and its performance ultimately. For a small number of inputs and a simple function, even a single hidden layer with a couple of neurons could do the job, however when solving complex functions, using a small numbers of neurons in the hidden layer is not enough and will lead the model to under-fit. On the flip side, using a very large amount of hidden neurons or hidden layers may cause the model to overfit and increase its training time.

In general, there is no fixed rule of thumb to determine the optimal number of hidden units and layers in a network and the decision depends upon trial and error. Some researchers prefer to adopt the forward building approach of designing a network, where they take a start from a small model and slowly and gradually increase the number of neurons until the training and test error gets improved and generalisation is achieved. In contrast, to achieve the same goal of determining the optimal number of hidden units in each hidden layer, a *pruning approach* is devised. Pruning involves the idea of removing nodes from the network during training and identifying those neurons whose removal does not noticeably affect the network's performance. In order to implement pruning, many researchers start with a large network configuration and through regularisation techniques prune the model such that the weights of removable hidden neurons become close to zero [10, 11]. Besides these techniques, some researchers advocate the following rules to select the appropriate number of hidden units:

- The amount of hidden units should between size the output layer and input layer [12].
- The amount of hidden units should not exceed to twice the size of the input units [13].
- The amount of specified hidden nodes should capture 70−90% of the variance on the input data set [14].

3.1.4   Initializing Weights of Deep models

It is recommended to initialize the weights of a deep model with smaller values. Large random values might speed the learning rate but end-up with worse feature representations. It is important to choose smaller weight values such that there is minimal or no uniformity between different unis. Typically, the weights are initialised from zero mean Gaussian distribution having a standard deviation of 0.01.

3.1.5   Learning Rate

Learning rate is one of the most important parameters in deep models as it manages the pace at which network parameters are optimised. It is usually a number chosen in the range [0, 1]. If the learning rate is too small, network takes a large amount of time to converge. On the contrary, if its too large then the weights of the networks are more likely to explode leading to fluctuation around the minima or divergence. To avoid both the extremes, one can try several mid range values and see the impact of learning on the cost function using the validation set. The learning rate that returns smallest cost on the cross validated set should be retained for the experiment at hand.

Instead of using a fixed learning rate, we can also opt for an adaptive learning rate that calls for a large learning rate initially and then its value is decreased gradually after several epochs. Using an adaptive learning rate might help us train the model faster as compared to the fix learning rate, however the decision of when to change the rate needs to be carefully taken manually.

If we do not want the same learning rate to be applied to all parameter updates, we can also apply some other sophisticated techniques of choosing learning rates such as Adagrad, Adadelta, RMSProp and Nadam [15]. If the input data is sparse, using one of these adaptive learning rate schemes can free us from the trouble of tuning a fixed learning rate manually.

3.1.6 The Size of Mini-Batch and Stochastic Learning

Rather than updating the network weights after the gradient estimation on a single training example as done in stochastic learning, it is more efficient to divide the training set into "mini-batches" of size 10–100 examples and then proceed with gradient estimation on each mini-batch. This strategy is considered favourable to take advantage of the parallel architecture on which implementation is being executed. The mini batch size is often taken as a power of two (32, 64, 128, 256, etc.) so that it fits the memory requirements of the GPU and CPU. It is generally recommended to tune the batch size and learning rate after all the other hyper-parameters are tuned [16].

3.1.7 Regularisation Parameter

Adding regularisation parameter in any deep architecture will help the deep models to prevent overfitting and increase the performance of the model. Some notable algorithms used to implement regularisation are swapout [17], dropout [18] and stochastic depth [19] algorithms.

3.1.8 Number of Iterations of Gradient Based Algorithms

Training a model with maximum number of epochs generally gives the model more room to reduce its loss function, however this training comes at a high computational cost and time expense. In order to mitigate this computational cost, a straight forward rule is applied, i.e.: Train the model on a fixed number of examples and epochs. After training each example, compare the test error with the train error. If the gap between train and test error is decreasing, then keep training, otherwise change the number of iterations.

- Visualize the training process after each epoch by plotting the histograms of weights and reconstruction error. It helps one keep track of whether the model is learning correctly or not.
- If the order of training examples (in different batch sizes or in different epochs) is shuffled, you will notice a slight boost in the convergence of the model.

- Keep the weights of the dimensions as exponential power of 2. It will boost the learning efficiency of your model by sharding the weights and matrices etc., if you keep the size of your hyper-parameters as 64, 128 etc.

3.1.9 Parameter Tuning: Evade Grid Search—Embrace Random Search

Although grid search has been pervasive in classical machine learning, it cannot prove efficient in finding optimum parameters for ultra deep models due to its computational complexity. In deep architectures, when the network size grows larger and the number of parameters increase massively, the computational cost of doing grid search for each one of them increases exponentially. In such situations, instead of using grid search, random sampling approach is utilised for parameter selection. In this approach, the combination of hyper-parameters are chosen through uniform distribution within a certain range. In practice, it is also useful to use prior information such as learning rate to further decrease the search space for finding optimal parameters.

3.2 Constructing Fisher Kernels from Deep Models

Once a generative model is trained, we next seek to learn how the model maximises its log likelihood to reproduce the seen data and reduces its cost. The Fisher encodings used in Fisher kernel are a buy product of the training mechanism that the model has used. In the following sections, we illustrate how Fisher kernels could be drawn from restricted Boltzmann machine and deep Boltzmann machine.

3.2.1 Demonstration of Fisher Kernel Extraction from Restricted Boltzmann Machine (RBM)

As an illustration, lets consider the example of drawing Fisher kernel from restricted Boltzmann machine (RBM), a stochastic generative model from the family of neural networks consisting of a bipartite graph fully connecting the visible and hidden units through undirected weight connections. There are no visible-visible and hidden-hidden connections, however each unit in the network is connected to an external bias unit. See Fig. 3.1 to comprehend the architecture of the generative model. The energy of the joint configuration of visible and hidden units is given as:

$$E(\mathbf{v}, \mathbf{h}; \boldsymbol{\theta}) = -\sum_{i=1}^{V}\sum_{j=1}^{H} w_{ij} v_i h_j - \sum_{i=1}^{V} b_i v_i - \sum_{j=1}^{H} a_j h_j, \tag{3.1}$$

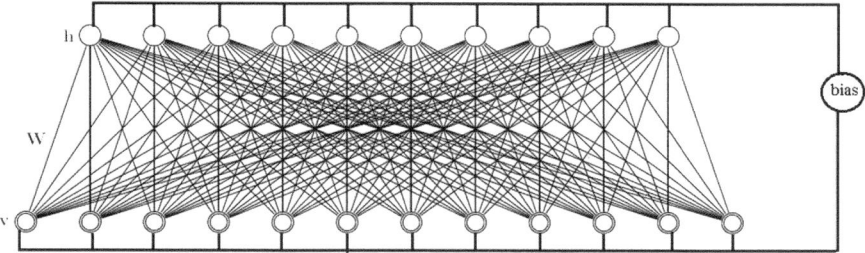

Fig. 3.1 A restricted Boltzmann machine composed of stochastic binary units with symmetric connections. The top layer represents the hidden units **h** and the bottom layer represents the visible units **v**. The weight vector W determines the connections between the units in the two layers

where v_i and h_j are the binary states of visible unit i and hidden unit j, w_{ij} represents the symmetric connection between the visible unit i and hidden unit j, b_i and a_j are the respective bias terms for visible and hidden units and $\theta = \{W, \mathbf{b}, \mathbf{a}\}$. The probability of a joint configuration over both visible and hidden units depends on the energy of this joint configuration compared with the energy of all other joint configurations:

$$P(\mathbf{v}, \mathbf{h}; \theta) = \frac{1}{Z(\theta)} \exp\left(-\mathbf{E}(\mathbf{v}, \mathbf{h}; \theta)\right), \tag{3.2}$$

where $Z(\theta)$ is known as the partition function or the normalisation constant, mathematically defined as:

$$Z(\theta) = \sum_{\mathbf{v}} \sum_{\mathbf{h}} \exp\left(-E(\mathbf{v}, \mathbf{h}; \theta)\right). \tag{3.3}$$

The parameters of this energy-based model, $\theta = \{W, \mathbf{a}, \mathbf{b}\}$ are learnt by performing (stochastic) gradient descent learning on the empirical negative log-likelihood $\ell(\theta, \mathcal{D})$ of the training data using the stochastic gradient $\nabla_\theta \log P(\mathbf{v}; \theta)$.

Mathematically $\log P(\mathbf{v}) = \log \sum_{\mathbf{h}} \exp(-E(\mathbf{v}, \mathbf{h})) - \log Z(\theta)$.

The gradient of the log likelihood of the observations, $P(\mathbf{v}; \theta)$ with respect to each model parameter $\theta = \{W, \mathbf{a}, \mathbf{b}\}$ is given as below:

$$\frac{\partial \log P(\mathbf{v}; \theta)}{\partial W} = \langle \mathbf{v}\mathbf{h}^T \rangle_{P_{data}} - \langle \mathbf{v}\mathbf{h}^T \rangle_{P_{model}}, \tag{3.4}$$

$$\frac{\partial \log P(\mathbf{v}; \theta)}{\partial \mathbf{a}} = \langle \mathbf{h} \rangle_{P_{data}} - \langle \mathbf{h} \rangle_{P_{model}}, \tag{3.5}$$

$$\frac{\partial \log P(\mathbf{v}; \theta)}{\partial \mathbf{b}} = \langle \mathbf{v} \rangle_{P_{data}} - \langle \mathbf{v} \rangle_{P_{model}}. \tag{3.6}$$

Here $\langle . \rangle_{P_{data}}$ denotes an expectation with respect to the data distribution $P(\mathbf{h}|\mathbf{v})$ and $\langle \ \rangle_{P_{model}}$ denotes an expectation with respect to the model distribution $P(\mathbf{v}, \mathbf{h})$. and is calculated approximately using *Contrastive Divergence* due to its algebraic intractability [20]. The Fisher Score which is the gradient of the log-likelihood function and transforms the sample input $\mathbf{x}$ with variable length into a fixed length vector depending on the number of parameters in the model, becomes as follows:

$$\nabla_\theta \log P(\mathbf{v}; \boldsymbol{\theta}) = \left[\frac{\partial \log P(\mathbf{v}; \boldsymbol{\theta})}{\partial W} \Big| \frac{\partial \log P(\mathbf{v}; \boldsymbol{\theta})}{\partial \mathbf{a}} \Big| \frac{\partial \log P(\mathbf{v}; \boldsymbol{\theta})}{\partial \mathbf{b}} \right]. \qquad (3.7)$$

3.2.2 MATLAB Implementation of Fisher Kernel Derived from Restricted Boltzmann Machine (RBM)

In order to derive a Fisher kernel from RBM, the generative model needs to be trained first so that the kernel could be derived. In this regard, we take a start by splitting our data set into train and test groups so that the model learns its parameters from the train data set only.

```
%Split Dataset into Train and Test Set
[train_data,test_data,train_labels,test_labels]=split(data,labels);
```

3.2.2.1 Train RBM Generative Model

After data-set splitting, one can learn optimal parameters of the RBM model using the train subset. The model parameters include the weights between the visible and hidden layers and bias connections connected to visible and hidden layers. Training RBM requires the use of following optimal hyper-parameters: learning rate, batch-size, number of hidden units and momentum. For choosing appropriate values of these hyper-parameters, see Sect. 3.1 for the guidelines.

```
%Train RBM model on Train Features
option.learningrate=0.01;
option.momentum=0.9;
option.batchsize=100;
numhid=4;
[model, errors] = rbmFit(train_data, numhid, train_labels,option);
```

3.2.2.2 Compute Fisher Scores

Once the generative model is trained, we can proceed to compute the Fisher Scores for each example in the given data set.

```
%Compute Gradient of Log-likelihood
[train_Fisherscore]=compute_Fisherscore(model,train_data);
[test_Fisherscore]=compute_Fisherscore(model,test_data);
```

3.2.2.3   Fisher Vector Normalisation

It is generally suggested to normalise the Fisher vectors for better classification performance. Before computing the Fisher kernel, it is suggested to normalise the Fisher scores through Min-Max normalisation approach discussed in Sect. 2.2.1.

```
%Fisherscore Normalization
Option='min-max';% You can also perform L1/L2.....
%norm and power normalization by changing the option value
[normalized_trainfeatures,normalized_testfeatures]=FV_normalize(train_Fisherscore,...
test_Fisherscore,option);
```

3.2.2.4   Compute Fisher Kernel

Now we can compute the Fisher Kernel on the normalised Fisher scores by using the commands below:

```
%Compute Fisher Kernel
train_Fishervector =normalized_trainfeatures' *normalized_trainfeatures;
test_Fishervector =normalized_testfeatures' *normalized_testfeatures;
```

3.2.2.5   Train Support Vector Machines with Fisher Kernel

After the formulation of Fisher kernel derived from RBM, the kernel could be embedded into any discriminative classifier like support vector machines (SVM) for performance evaluation. The following code snippet sets the regularisation parameter, C of SVM at 2^0, however this needs to be finalised with the assistance of grid search method. Grid search method is an exhaustive parameter selection approach particularly expensive in case of large data sets where it takes a lot of time to find the best parameter setting. Alternative parameter selection methods [21, 22] are proposed to optimise this parameter for SVM in case of large data sets or resource constrained environments.

```
%Support Vector Machine Classsifier
option.kernel_type =0; %Type of kernel, linear:0; rbf:2
option.c = 2^0; %Regularization parameter 'C' in SVM training
type = sprintf('-s 0 -t 0 -c %f',option.kernel_type,option.c);
% Train SVM with Fisher Kernel
[model_svm]=svmtrain(train_labels,train_Fishervector,type);
[predicted_labels]= svmpredict(test_labels,test_Fishervector,model_svm);
cp=classperf(predicted_labels,test_labels);
cp.CorrectRate;%Accuracy rate
```

3.2.3 Illustration of Fisher Kernel Extraction from Deep Boltzmann Machine

Deep Boltzmann machine is a multi-layered generative model capable of learning complex patterns in data through multiple hidden layers. Unlike deep belief networks, the model is completely un-directed and the connections only exist between the units of neighbouring layers. For a 3-layered deep Boltzmann machine, the joint configuration of visible and hidden units are expressed as the following energy function:

$$E(\mathbf{v}, \mathbf{h}; \boldsymbol{\theta}) = \mathbf{v}^T \mathbf{W}^1 \mathbf{h}^1 + \mathbf{h}^1 \mathbf{W}^2 \mathbf{h}^2 + \mathbf{h}^2 \mathbf{W}^3 \mathbf{h}^3, \tag{3.8}$$

where $\mathbf{h} = \{\mathbf{h}^1, \mathbf{h}^2, \mathbf{h}^3\}$ are the set of hidden units in each respective layer, and $\boldsymbol{\theta} = \{\mathbf{W}^1, \mathbf{W}^2, \mathbf{W}^3\}$ are the model parameters, representing visible-to-hidden and hidden-to-hidden symmetric weight connections. The probability that the model assigns to each visible vector $\mathbf{v}$ is given as:

$$P(\mathbf{v}; \boldsymbol{\theta}) = \frac{P^*(\mathbf{v}; \boldsymbol{\theta})}{Z(\boldsymbol{\theta})} = \frac{1}{Z(\boldsymbol{\theta})} \sum_{\mathbf{h}} \exp(-E(\mathbf{v}, \mathbf{h}^1, \mathbf{h}^2, \mathbf{h}^3; \boldsymbol{\theta})).) \tag{3.9}$$

The derivative of the log-likelihood with respect to model parameters $\boldsymbol{\theta}$ takes the following form:

$$\frac{\partial \log P(\mathbf{v}; \boldsymbol{\theta})}{\partial \boldsymbol{\theta}} = E_{P_{data}}[\mathbf{v}\mathbf{h}^T] - E_{P_{model}}[\mathbf{v}\mathbf{h}^T], \tag{3.10}$$

where $E_{P_{data}}[.]$ denotes an expectation with respect to the complete data distribution $P_{data}(\mathbf{h}, \mathbf{v}; \boldsymbol{\theta}) = \mathbf{P}(\mathbf{h}|\mathbf{v}; \boldsymbol{\theta}) \, P_{data}(\mathbf{v})$, with $P_{data}(\mathbf{v}) = \frac{1}{n} \sum_n \delta(\mathbf{v} - \mathbf{v}_n)$ representing the empirical distribution and $E_{P_{model}}[.]$ denoting an expectation with respect to the distribution defined by the model. The exact computation of data dependent distribution is exponential in the number of hidden units, whereas the model's expectation is exponential in the number of hidden and visible units. For this very reason, the exact maximum likelihood learning in DBM is intractable and is carried out via approximation schemes and variational approaches. The convention has been to use *mean-field inference*, a variational approach to estimate data dependent expectations, $E_{P_{data}}[.]$ and *Markov Chain Monte Carlo (MCMC)* based stochastic approximation procedure to approximate the model's expected sufficient statistics, $E_{P_{model}}[.]$ [23]. The Fisher score $\phi_{\mathbf{v}}$ is derived from the log likelihood of this generative model with respect to its parameters $\boldsymbol{\theta} = \{\mathbf{W}^1, \mathbf{W}^2, \mathbf{W}^3\}$ as given below:

$$\phi_{\mathbf{v}} = \nabla_{\boldsymbol{\theta}} \log P(\mathbf{v_n}|\boldsymbol{\theta}) = \left[\mathbf{S}_{[n]} \big| \mathbf{Q}_{[n]} \big| \mathbf{U}_{[n]} \right], \text{ where}$$

$$\mathbf{S}_{[n]} = \nabla_{\mathbf{W}^1} \log P(\mathbf{v_n}|\mathbf{W}^1) = \langle \mathbf{v}\mathbf{h}^{1^T} \rangle_{P_{data}} - \langle \mathbf{v}\mathbf{h}^{1^T} \rangle_{P_{model}},$$

$$\mathbf{Q}_{[n]} = \nabla_{\mathbf{W}^2} \log P(\mathbf{v_n}|\mathbf{W}^2) = \langle \mathbf{h}^1\mathbf{h}^{2^T} \rangle_{P_{data}} - \langle \mathbf{h}^1\mathbf{h}^{2^T} \rangle_{P_{model}},$$

$$\mathbf{U}_{[n]} = \nabla_{\mathbf{W}^3} \log P(\mathbf{v_n}|\mathbf{W}^3) = \langle \mathbf{h}^2\mathbf{h}^{3^T} \rangle_{P_{data}} - \langle \mathbf{h}^2\mathbf{h}^{3^T} \rangle_{P_{model}}.$$

The angle brackets, $\langle . \rangle$ in the above equations refer to the expected value over a certain distribution specified by the subscript it follows: P_{data} refers to the probability distribution $P(\mathbf{h}|\mathbf{v})$ which is tractable, whereas P_{model} refers to the probability distribution, $P(\mathbf{v}, \mathbf{h})$, which could not be calculated analytically as the model becomes excessively large with many hidden units. In practice, this learning is accomplished by following an approximation to the gradient of a different objective function, called the *Contrastive Divergence* [20].

3.2.4 MATLAB Implementation of Fisher Kernel Derived from Deep Boltzmann Machine (DBM)

Train Deep Boltzmann Machine

For deriving a Fisher kernel from deep Bolztmann machine (DBM), the generative model is trained first on the training set.

```
% Training First Layer of DBM %
num_hidden_units=100; max_epoch=30;
restart=1;
rbm;
%Training Second Layer of DBM %
num_hidden2 = 1000; maxe_poch=200;
restart=1;
rbm_12;
%Learning and Training of Two-Layer Deep Boltzmann Machine %
num_hidden_units=100; num_hidden2 = 1000;max_epoch=200;
restart=1;
makebatches;
dbm_mf;
%Save Visible/Hidden Units Learned Values%
save dbm labpen labbiases hidpen penbiases vishid hidbiases visbiases;
%Perform Fine-tuning Of Two-Layer Deep Boltzmann machine %
maxepoch=100;
makebatches;
backprop
%Create Structure of Learned Parameters
 model=struct('labpen',labpen,'labbiases',labbiases,'penbiases',.....
 penbiases,'vishid',vishid,'hidbiases',hidbiases,'visbiases',visbiases);
```

Compute Fisher Scores

Once the generative model is trained, we can proceed to compute the Fisher scores for each example in the given data set. For any n examples, we shall be able to collect n Fisher scores.

```
%Compute Gradient of Log-likelihood
[train_Fisherscore]=compute_Fisherscore(model,train_data);
[test_Fisherscore]=compute_Fisherscore(model,test_data);
```

Fisher Vector Normalisation

Next, we normalise the Fisher vectors to achieve better classification/retrieval performance.

```
%Fisherscore Normalization
Option='min-max';% You can also perform L1/L2.....
%norm and power normalization by changing the option value
[normalized_trainfeatures,normalized_testfeatures]=FV_normalize(train_Fisherscore,...
test_Fisherscore,option);
```

Compute Fisher Kernel

The Fisher Kernel is computed from the normalised Fisher scores.

```
%Compute Fisher Kernel
train_Fishervector =normalized_trainfeatures' *normalized_trainfeatures;
test_Fishervector =normalized_testfeatures' *normalized_testfeatures;
```

Train SVM with Fisher Kernel

After the formulation of Fisher kernel derived from DBM, the kernel could be embedded into any discriminative classifier like support vector machines for performance evaluation.

```
%Support Vector Machine Classsifier
option.kernel_type =0; %Type of kernel, linear:0; rbf:2
option.c = 2^0; %Regularization parameter 'C' in SVM training
type = sprintf('-s 0 -t 0 -c %f',option.kernel_type,option.c);
% Train SVM with Fisher Kernel
[model_svm]=svmtrain(train_labels,train_Fishervector,type);
[predicted_labels]= svmpredict(test_labels,test_Fishervector,model_svm);
cp=classperf(predicted_labels,test_labels);
cp.CorrectRate;%Accuracy rate
```

References

1. LeCun, Y., Bottou, L., Orr, G.B., et al.: Efficient backprop. In: Neural Networks: Tricks of the Trade, pp. 9–50. Springer (1998)
2. Bengio, Y.: Practical recommendations for gradient-based training of deep architectures. In: Neural Networks: Tricks of the Trade, pp. 437–478. Springer (2012)
3. Marchesi, M.: Megapixel size image creation using generative adversarial networks (2017). arXiv preprint arXiv:1706.00082
4. Wang, J., Perez, L.: The effectiveness of data augmentation in image classification using deep learning (2017). arXiv preprint arXiv:1712.04621

5. Ahmed, S., Azim, T.: Compression techniques for deep fisher vectors. In: ICPRAM, pp. 217–224 (2017)
6. Collobert, R., Bengio, S.: Links between perceptrons, MLPs and SVMs. In: Proceedings of the Twenty-First International Conference on Machine Learning. ACM (2004)
7. Jarrett, K., Kavukcuoglu, K., LeCun, Y., et al.: What is the best multi-stage architecture for object recognition? In: IEEE 12th International Conference on Computer Vision, pp. 2146–2153. IEEE (2009)
8. Glorot, X., Bengio, Y.: Understanding the difficulty of training deep feedforward neural networks. In: Proceedings of the Thirteenth International Conference on Artificial Intelligence and Statistics, pp. 249–256 (2010)
9. Glorot, X., Bordes, A., Bengio, Y.: Deep sparse rectifier neural networks. In: Proceedings of the Fourteenth International Conference on Artificial Intelligence and Statistics, pp. 315–323 (2011)
10. Louizos, C., Welling, M., Kingma, D.: Learning sparse neural networks through L_0 regularization (2017). arXiv preprint arXiv:1712.01312
11. Theis, L., Korshunova, I., Tejani, A., et al.: Faster gaze prediction with dense networks and fisher pruning (2018). arXiv preprint arXiv:1801.05787
12. Blum, A.: Neural Networks in C++, vol. 697. Wiley, NY (1992)
13. Berry, M., Linoff, G.: Data Mining Techniques: For Marketing, Sales, and Customer Support. Wiley (1997)
14. Boger, Z., Guterman, H.: Knowledge extraction from artificial neural network models. In: IEEE International Conference on Systems, Man, and Cybernetics Computational Cybernetics and Simulation, vol. 4, pp. 3030–3035. IEEE (1997)
15. Ruder, S.: An overview of gradient descent optimization algorithms. Comput. Res. Repos. (2016). http://arxiv.org/abs/1609.04747. (CoRR) abs/1609.04747
16. Bengio, Y.: Practical recommendations for gradient-based training of deep architectures. Comput. Res. Repos. (2012). CoRR abs/1206.5533
17. Singh, S., Hoiem, D., Forsyth, D.: Swapout: learning an ensemble of deep architectures. In: Advances in Neural Information Processing Systems, pp. 28–36 (2016)
18. Srivastava, N., Hinton, G., Krizhevsky, A., et al.: Dropout: a simple way to prevent neural networks from overfitting. J. Mach. Learn. Res., 1929–1958 (2014)
19. Huang, G., Sun, Y., Liu, Z., et al.: Deep networks with stochastic depth. In: European Conference on Computer Vision, pp. 646–661. Springer (2016)
20. Hinton, G.: Training products of experts by minimizing contrastive divergence. Neural Comput., 1771–1800 (2002)
21. Lin, J., Zhang, J.: A fast parameters selection method of support vector machine based on coarse grid search and pattern search. In: Fourth Global Congress on Intelligent Systems, pp. 77–81 (2013)
22. Staelin, C.: Parameter selection for support vector machines. Technical report, Hewlett-Packard Company, HPL-2002-354R1 (2003)
23. Salakhutdinov, R., Larochelle, H.: Efficient learning of deep Boltzmann machines. In: Proceedings of the Thirteenth International Conference on Artificial Intelligence and Statistics, pp. 693–700 (2010)

Chapter 4
Large Scale Image Retrieval and Its Challenges

Abstract Fisher vector encoding from deep architectures has shown significant improvement in the performance of classification and retrieval tasks. Despite significant benefit of Fisher vectors for classification and retrieval problems, they suffer from the problem of high dimensionality giving rise to computational and storage overhead for large scale learning problems. This chapter provides guidelines for tackling this issue by either deploying feature selection or compression methods. We provide an overview of all the popular feature selection and compression techniques and identify some metrics that can help practitioners identify the appropriateness of each technique used for the cause.

Keywords Large scale image retrieval · Feature selection · Feature compression
Multi-collinearity · Variance ination factor (VIF) · Filters · Wrappers
Minimum redundancy Maximum relevance (MRMR) · Parametric t-SNE

With the emergence of world wide web and spread of mobile computing, the amount of digital data has increased massively giving rise to large scale learning problems in image retrieval and text mining. This new era of digital age demands scalable algorithms that could easily be deployed in the wake of current challenges and address the growing demands of the society. To cope with these challenges, the researchers have dedicated a lot of attention to learn image representations for large scale image categorisation and retrieval problem. One of the successful approaches for the task is to obtain local features [1] from the image and encode them as high dimensional vectors. Till date, various encoding methods are proposed for this purpose, such as sparse coding [2], bag of visual words [3, 4] and Fisher Vector encoding [5]. In comparison to other encoding methods, Fisher vectors have consistently shown outstanding performance on different image recognition tasks [6, 7]. The supremacy of Fisher kernels for object classification task was first shown by Holub et al. [8] on Caltech data set who successfully combined the probabilistic constellation model with kernel methods. Following them, Perronin and Dance [9] applied the Fisher kernel framework to a visual vocabulary of low-level feature vectors extracted from images and modelled them via the Gaussian mixture model (GMM). The authors showed that the proposed approach acts as a generalisation of popular bag of words (BoW) approach and makes the image representation more informative even if the available vocabulary is limited. This seminal work opened up a new avenue of classification experiments on many large scale object recognition data sets such as CalTech-256, PASCAL VOC 2007 and Imagenet LSVRC [10–13]. The supremacy of kernel methods was overtaken by deep models in 2012 when Krizhevsky et al. [14] showed that AlexNet, a variant of convolution neural network (CNN) can achieve a top-5 test error rate on classification problem of ImageNet. Since this breakthrough, various improved versions of deep models have consistently shown remarkable improvement over state of the art performances in various computer vision and machine learning problems. While deep models have maintained a performance lead, its important to note that their supremacy is indebted to the availability of large amount of data, sophisticated hardware and improved optimisation techniques. For any of the problem comain, where the computational requirements of the deep models could not be met, the proposed deep learning solutions cannot help. Keeping this in view, the research community has recently proposed to use hybrid strategies combining deep models with kernels to bring the best out of both the paradigms [15–18] in terms of performance and computational overhead. Many of these results suggest that Fisher kernels match deep models and in some cases may surpass the performance of the deep models.

The standard practice of deriving Fisher vectors from deep models faces two challenges: (i) Results in large dimensional Fisher vectors that require more memory usage and disk space for classifier training, (ii) Prone to overfitting under limited data availability applications due to the large dimensions of Fisher scores extracted from deep models. In the sections below, we shall discuss how to cope with these constraints and make them amenable for large-scale image retrieval and classification applications.

4.1 Condensing Deep Fisher Vectors: To Choose or to Compress?

Most of the measured features in high dimensional Fisher vectors do not contribute in understanding the underlying structure of the data and hence these dimensions need to be reduced carefully to improve storage and computational cost as well as yield better model performance. *Feature compression* and *feature selection* are the two popular off the shelf dimensionality reduction techniques in practice for reducing data's high dimensional memory footprint and increase its suitability for large scale visual retrieval and classification. There is a dire need to have a metric for guiding our selection of feature condensation methods. According to [19–21] the existence of high *multi-collinearity* among the dimensions of Fisher vectors indicates that feature compression is a better choice than feature selection as it transforms the features from original space to a new space and minimizes the effect of multi-collinearity among the features. While feature selection methods choose the subset of features from the original space; in case of high multi-collinearity, there might be a possibility that we would eliminate those features that have a strong impact on other features and on model performance. So dropping/choosing features may degrade the accuracy of the model [21].

4.2 How to Detect Multi-collinearity?

Multi-collinearity exists when there is a high correlation among the dimensions of the data set. The linear relationship among more than one dimensions in the data results in redundancy leading to poor model performance. There are different statistical measures to assess multi-collinearity, i.e. Pearson correlation, variance inflation factor and condition indices. In this section, we only discuss variance inflation factor for multi-collinearity assessment.

4.2.1 Variance Inflation Factor (VIF)

Variance inflation factor shows how much is the variance (standard error) inflated due to the existence of correlation between independent variables in a model. This may not necessarily be a problem, but it can prevent precise analysis of the individual effects of each variable. Mathematically, it is defined as the reciprocal of *tolerance* given as $1 - R_i^2$, where R_i corresponds to the value predicted by regressing i-th variable by the rest of independent variables. A tolerance close to 1 means that there is little multi-collinearity, whereas a value close to 0 suggests that multi- collinearity may be a threat. Conversely, the variance inflation factor is elaborated by the following piecewise function:

$$VIF \begin{cases} \approx 0 & \implies \text{Moderate to null multi-collinearity} \\ > 5 & \implies \text{High multi-collinearity} \end{cases}$$

Conventionally, the threshold used for large VIF values is 5, however some of the literature also uses 10 as a threshold, i.e. 0.10 tolerance factor to indicate multi-collinearity among independent variables.

4.2.1.1 Matlab Implementation of Variance Inflation Factor

Let `train_Fishervector` denotes the Fisher score derived from any deep model having n parameters and thus n dimensions, i.e. $D = [D_1........D_n]$. In order to compute variance inflation factor (VIF), one has to identify the dependencies among features by giving the following commands below:

```
%Identify Dependencies among Features
r=corrcoef(train_Fishervector);
%Calculation of VIF
features_vifscores=diag(inv(r))';
```

The code shall return an n dimensional column vector, $[1 \times n]$ stored in the variable named as `features_vifscores`, specifying the variance inflation factor score of each dimension.

4.3 Feature Compression Methods

Feature compression is the process of transforming high dimensional data with dimensionality D into a low dimensional space d, where $d < D$. The features are compressed in such a way that the global and local structure of the data is preserved as much as possible. Over the last few decades, different feature compression techniques are proposed for visualizing high-dimensional data [22–25]. In this section, we will give an overview of various popular feature compression techniques.

4.3.1 Linear Feature Compression Methods

Linear techniques project data from high dimensional feature space into a low dimensional feature space using linear projections. The solution space of linear techniques is convex which means that the objective function of such methods does not contain any local optima. Most of the feature compression methods fall in the category of convex techniques. Some of the popular linear techniques for feature compression are principal component analysis (PCA) [26], classical scaling [27], linear discriminant anlaysis (LDA) and many more. In these techniques, the objective function is in the form of Rayleigh quotient and can be optimized by solving eigen-value decomposition problem. These approaches compute the pairwise similarities between the data points and capture the covariances between feature dimensions by performing eigen-decomposition of a full matrix. Linear compression methods have the inability to embed non-linear data well into latent space.

4.3.2 Non-linear Feature Compression Methods

Non-linear feature compression methods preserve the global structure of the data by constructing non-linear representation of data in low dimensional latent space. Some of the non-linear compression techniques include multi-dimensional scaling (MDS), ISOMAP, stochastic neighbor embedding (SNE) and auto-encoder, etc. Mutidimensional scaling retains the pairwise distance between data-points by mapping high dimensional data into low dimensional space. This mapping can be modeled using a stress function. While auto-encoder is a multi-layer neural network that can learn non-linear mapping by maximizing the variance of the data in low-dimensional space. The local non-linear feature compression techniques retains the local structure of small neighborhoods around the data points. The local non-linear feature compression methods are local linear embedding (LLE) and Laplacian eigenmaps etc.

4.4 Feature Selection Methods

Feature selection techniques aim to select a subset of features from the original space without doing any manifold transformation. It aims to maintain the representation of features in its original space and selects minimum number of significant feature having an impact on the performance of the application. Selecting the best features from a data set is a NP-complete problem. The task is considered challenging because of two reasons: First, the features that do not appear significant alone may be extremely relevant when taken in combination with other features. Secondly, there is a chance that relevant features in a group are redundant and are therefore responsible for unnecessary complexity that may lead to overfitting later. An exhaustive search of all possible feature subsets ensures the existence of best feature subset that contains the least number of features contributing towards accuracy. However, this is not a computationally feasible approach and is often replaced by heuristic and meta-heuristic searches that reduce the search space for feature selection. Feature selection techniques are broadly categorized as filters, wrappers, and embedded methods introduced briefly below:

4.4.1 Feature Selection via Filter Methods

Filters evaluate the relevancy of features by looking at the underlying properties of the data set. It ranks the features in order of their relevance score and eliminates those features having a low relevance scoring. To accomplish the task, once the features are ranked, a subset of **N** best features are chosen and given to the classification algorithm. Filter techniques can either rank individual features by analyzing the pair-wise dependency of each individual feature or by evaluating the whole feature subset. There are several statistical criteria for the evaluation of feature dependency, such as correlation, hypothesis test, uncertainty and discriminative power. Some of the most popular filter methods are chi-squared test, information gain [28], relief and its variants [29, 30], mutual information [31] and maximum relevance minimum redundancy (MRMR) [32], etc. Filter methods are computationally fast, simple to implement, scalable and are independent of the deployed classification algorithm. Their ignorance of the used classifier often affects their contribution towards accuracy.

4.4.2 Feature Selection via Wrapper Methods

In wrapper-based feature selection approach, a subset of features is chosen using a search strategy that evaluates the goodness of each feature subset through classifier performance and keeps searching iteratively for a subset of features until the desired performance is achieved. This approach wraps up a classifier in a feature selection procedure and acts as black box in which the optimal features depend upon the performance evaluation criterion of the induction algorithm [33, 34]. Finding optimal features through wrapper methods is a combinatorial problem and needs *complete search strategy* in-order to locate the best possible feature subset. This strategy becomes computationally expensive as the number of features become very large and is therefore replaced by *heuristic* search, *meta-heuristic* search or methods that use *artificial neural networks*. Heuristics such as branch and bound algorithm reduce the search space as well as the time complexity of the complete search approach. However, the meta-heuristics can perform faster than heuristic and complete search methods due to their random nature. Examples of meta heuristic techniques include genetics algorithms, ant colony optimisation, etc. Generally, wrappers produce better results than filters due to their ability to model the relationship between the learning algorithm and the training data. On the flip side, wrappers are computationally expensive than the filters because for every chosen feature subset, the classifier must be executed repeatedly. Wrappers also have the tendency to overfit the solution.

4.4.3 Feature Selection via Embedded Methods

Embedded methods perform feature selection during the execution of a supervised learning algorithm. They are categorised into three different classes: (i) Built-in methods, (ii) Pruning methods and (iii) Regularisation methods. In the built-in techniques, classifier selects features as a part of the learning phase. Some common examples of *built-in* embedded methods include decision trees, logistic regression and its variants, random forests, etc. In *pruning based feature selection approach*, all the features are used to train the model first and then some features are removed by setting the corresponding feature co-efficients to zero. A popular example of this technique is recursive feature elimination method used in support vector machines (SVM). The last class of embedded methods uses *regularisation* to set or allocate different weight to features such that the error of the objective function is reduced.

The value of the regularisation term sets the importance of each feature and decides its level of contribution in the model performance. Regularisation parameters are an integral part of most of the state of the art machine learning algorithms and greatly assist in solving issues such as *overfitting* and *underfitting*.

4.5 Hands on Fisher Vector Condensation for Large Scale Data Retrieval

In this section, we provide guidelines to novice practitioners for condensing Fisher vectors used for large scale image data retrieval and classification applications. We will discuss two popular state of the art feature selection and compression techniques below: Maximum relevance and minimum redundancy (MRMR) and parametric t-distributed stochastic neighbor embedding (t-SNE).

4.5.1 Minimum Redundancy and Maximum Relevance (MRMR)

MRMR is a filter based selection approach [32] that ranks the features by maximising mutual information between the joint probability distribution of the selected features and the classification variable after calculating the minimum redundancy and maximum relevancy criterion. The algorithm penalises the relevancy $D(S, C)$ of features if redundancy $R(S)$ is present in this selection. The relevancy of feature set S for the class C is computed as:

$$D(S, C) = \frac{1}{|S|} \sum_{f_i \in s} I(f_i; c).$$ (4.1)

The redundancy of all features in the set S is estimated as:

$$R(S) = \frac{1}{|S|^2} \sum_{f_i, f_j \in s} I(f_i; f_j).$$ (4.2)

The feature selection criterion for MRMR is a combination of both these measures that maximises relevance and minimises redundancy as below:

$$MRMR = \max_{s} \left[\frac{1}{|S|} \sum_{f_i \in s} I(f_i; c) - \frac{1}{|S|^2} \sum_{f_i, f_j \in s} I(f_i; f_j) \right].$$ (4.3)

MRMR gives good accuracy with simple linear classifiers if the Fisher vectors encoding from deep models are normalised using min-max normalisation and entropy of mutual information is calculated using discretely quantized variable instead of computing probability density function [35, 36].

4.5.1.1   Matlab Implementation of MRMR

The code of MRMR is available on (http://home.penglab.com/proj/mRMR/).

Normalise Features

Before performing MRMR feature selection on Fisher vectors, its recommended to normalise the features first.

```
%Features Normalization
Option='min-max';% You ca also perform L1 norm, L2 norm ...
%and power normalization by changing the option value
[normalized_trainfeatures,normalized_testfeatures]=FV_normalize(train_features,...
test_features,option);
```

Perform Feature Selection

In this step you have to perform feature selection using MRMR scheme. There are generally two popular schemes to compute the mutual information: Mutual information difference (MID) and Mutual information quotient (MIQ). MID is the difference of the mutual information between features and the class label, whereas MIQ is the ratio of the mutual information between features and the class label.

```
%Perform Feature Selection
Param.method='miq';
[features_rank,features_Score]=ftSel_mrmr(normalized_trainfeatures,...
train_labels,param);
```

Selection of Required Number of Features from Train and Test Sets

After applying mRMR, the algorithm will rank all the features on the basis of their maximised relevance and minimised redundancy. One can choose required n features from both the training and test sets for further computation.

```
%Select Low Dimensional 'n' Features
dims=2;
for i=1:dims
    dimension=features_rank(:,i);
    X_train(:,i)=normalized_trainfeatures(:,dimension);
end
for i=1:dims
    dimension=features_rank(:,i);
    X_test(:,i)=normalized_testfeatures(:,dimension);
end
```

Perform Classification

In this step we compute the accuracy of our model using k-nearest neighbour with $k = 1$.

```
%Perform 1-Nearesr Neighbor Classifier
c=knnclassify(X_test,X_train,train_labels);
cp=classperf(c,test_labels);
cp.CorrectRate;%Accuracy rate
```

Useful Guideline for Using MRMR

As discussed above, MRMR uses two schemes for calculating the mutual information, i.e. MID and MIQ. According to the literature [37], both schemes give almost the same accuracies but MID is more stable when number of samples in a data set are smaller than the number of dimensions in the data set.

4.5.2 Parametric t-SNE

Parametric t-SNE [38] is an unsupervised dimensionality reduction technique that learns a parametric mapping between the high-dimensional and low-dimensional feature spaces such that the local structure of the data is preserved in manifold reduced space. In parametric t-SNE, the mapping $f : X \rightarrow Y$ from the high dimensional data space X to the low-dimensional latent space Y is parametrised by means of a feed-forward neural network with weights W. The training procedure is inspired by an auto-encoder based on restricted Boltzmann machine (RBM) that operates in three main stages: (1) First, a stack of RBMs is trained, (2) Next, the stack of RBMs is used to construct a pre-trained neural network, and (3) At the end, the pre-trained network is fine-tuned using back-propagation algorithm to minimise the cost function that retains local structure of the data in latent space by minimising the Kullback-Leibler (KL) divergence between the probabilities signifying pairwise distances between examples.

Impact of Tweaking Perplexity on Parametric t-SNE

Parametric t-SNE may behave mysteriously if the parameters are not tuned effectively. One of the most important parameters is called *perplexity* that can be viewed as a handle to adjust the performance of parametric t-SNE. Abstractly speaking, perplexity guesses the number of close effective nearest neighbours a point has. In order to get the most out of parametric t-SNE, one needs to analyse the visual plots in 2D/3D with different values of perplexity. The optimal value of perplexity typically ranges between 5 and 50 as suggested by Lauren and Hinton.

  We here walk through an example shown in Fig. 4.1 by changing the values of perplexity from 2–100, i.e. outside the prescribed range of [5–50]. In each of these

(a)   **(b)**

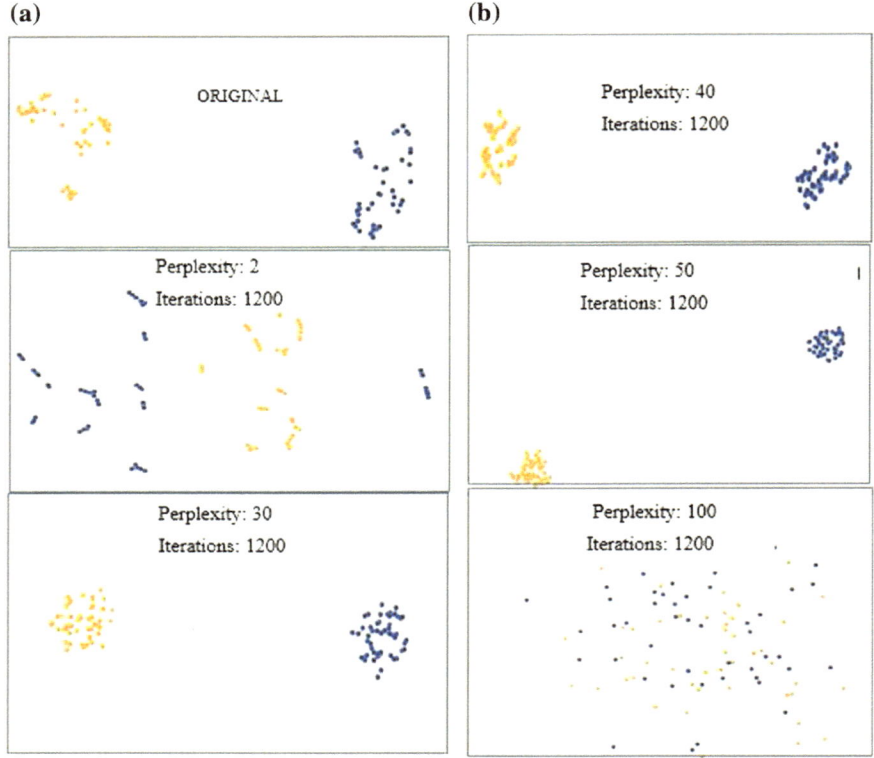

Fig. 4.2 Impact of tweaking perplexity on the performance of parametric t-SNE

plots, the total number of iterations were fixed at 1200 with learning rate ϵ taken as 10. When a small perplexity is chosen (less than 5), local variations in the data dominate whereas for perplexity greater than 50, parametric t-SNE shows undesirable results with merged clusters.

Impact of Number of Iterations on Parametric t-SNE

Another important parameter to get the best out of parametric t-SNE is the number of iterations. The general rule is to run the learning algorithm until it reaches a point of stability.

In the simulations shown in Fig. 4.2, one can observe that early stopping does not give significant results as the local structure of the data is not preserved. Moreover, the stability in results is not guaranteed by fixed number of iterations and different data sets require different number of iterations to converge. In such scenarios when early stopping does not help, it is necessary to increase the number of iterations instead of changing perplexity or other parameters of the algorithm. If the data-set on board is dense, then larger perplexity value gives better performance and allows the algorithm to converge.

(a) **(b)**

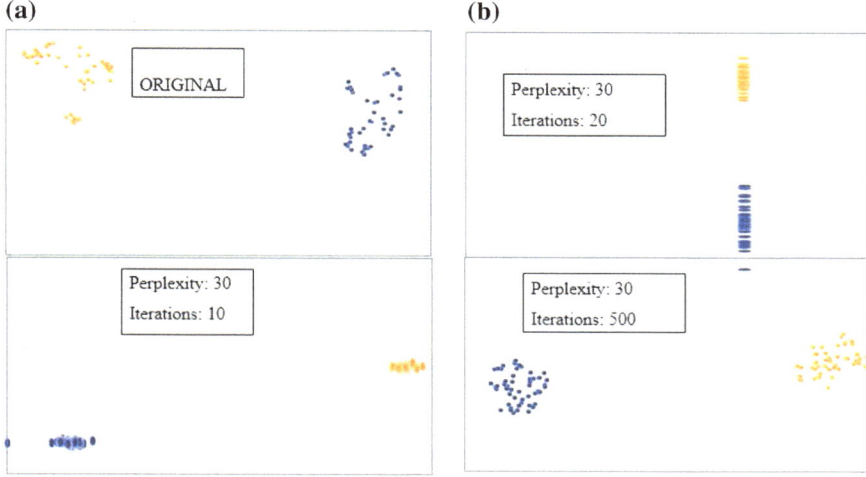

Fig. 4.2 Impact of changing the number of iterations in Parametric t-SNE

4.5.2.1 Additional Practical Guidelines to Use Parametric t-SNE Efficiently

- Parametric t-SNE outperforms the state of the art results if the Fisher vector encodings are normalised using min-max normalisation method [35]. Min-Max normalisation scheme re-scales the features in the range [0,1].
- The value of perplexity is usually taken in the range [5–50]. If the Fisher features are dense and dimensions are quite large, then the value of perplexity should not be less than 40.
- If the performance of parametric-tSNE does not improve with various learning rates and perplexity values, try increasing the number of iterations.
- Learning rate of 0.01 works well in majority of the cases. So you can take a start by setting the value of learning rate as 0.01 first and then explore other possible values if the results are not satisfactory.

4.5.2.2 Matlab Implementation of Parameteric t-SNE

The source code of parametric t-SNE could be downloaded from https://lvdmaaten. github.io/tsne/.

Parameter Settings

In order to condense Fisher vectors using parametric t-SNE, it is necessary to set optimal values for hyper-parameters utilised in training. These hyper-parameters include perplexity, number of iterations, layers and type of training algorithm. The source code offers two choices for training: Contrastive divergence-1 (CD-1) or persistent contrastive divergence (PCD). Both the training algorithms use an approximate method for sampling from the model distribution and have different variance and bias for the estimation of stochastic gradient data-points. Generally, CD-1 algorithm is faster and works better with smaller mini-batch size or with higher learning rate while PCD is also a good choice but it needs smaller learning rate which makes learning slow in-order to reduce the reconstruction error in RBM.

```
%Parameters Settings
perplexity = 30;
iterations=100;
layers = [1000 500 250 2];
training ='CD1';
```

Feature Normalisation

Before passing the data for feature compression, it is recommended to normalise the features (both training and test features) as a pre-processing step for attaining better performance. There are different choices for normalisation that have shown exceptional performances such as L1/L2 normalisation, min-max normalisation and power normalisation.

```
%Features Normalization
Option='min-max';% You ca also perform L1 norm, L2 norm ...
%and power normalization by changing the option value
[normalized_trainfeatures,normalized_testfeatures]=FV_normalize(train_features,...
test_features,option);
```

Train the Parametric t-SNE Network

Next, train the parametric t-SNE network that uses RBM and fine tunes the model using back-propagation algorithm for minimising the cost function that preserves the local structure of the data in reduced dimensional manifold. In order to tune RBM parameters, see the guidelines discussed in Chap. 3.

```
%Train Parametric t-SNE model on Features
[Network, error] = train_par_tsne(normalized_trainfeatures, train_labels,...
normalized_testfeatures, test_labels, layers, training);
```

Construction of Training and Test Embeddings Defined in Network

In this step, high dimensional data is embedded into a low dimensional space.

```
%Construction of  Training and Test Embeddings Defined in Network
mapped_train = run_data_through_network(network,normalized_trainfeatures);
mapped_test = run_data_through_network(network,normalized_testfeatures);
```

Computing Accuracy of Parmetric t-SNE Mapped Features Using a Classifier

For the sake of illustration, we compute the accuracy of our model using 1-nearest neighbour. Any other standard classifier could also be used instead of k-NN to check the discriminatory power of transformed data mappings.

```
%Apply 1-Nearest Neighbor Classifier
c=knnclassify(mapped_test,mapped_train,train_labels);
cp=classperf(c,test_labels);
cp.CorrectRate;%Accuracy rate
```

References

1. Lowe, D.: Object recognition from local scale-invariant features. International Conference on Computer Vision (ICCV) **2**, 1150–1157 (1999)
2. Yang, J., Yu, K., Gong, Y., Huang, T.: Linear spatial pyramid matching using sparse coding for image classification. In: IEEE Conference on Computer Vision and Pattern Recognition, pp. 1794–1801. IEEE (2009)
3. Csurka, G., Dance, C., Bray, C., et al.: Visual categorization with bags of keypoints. In: Workshop on Statistical Learning in Computer Vision (ECCV), pp. 1–22 (2004)
4. Sivic, J., Zisserman, A.: Video Google: A text retrieval approach to object matching in videos. In: null, p. 1470. IEEE (2003)
5. Perronnin, F., Sanchez, J., Mensink, T.: Improving the fisher kernel for large-scale image classification. In: European Conference on Computer Vision, pp. 143–156. Springer (2010)
6. Akata, Z., Perronnin, F., Harchaoui, Z., et al.: Good practice in large-scale learning for image classification. IEEE Trans. Pattern Anal. Mach. Intell. **3**, 507–520 (2014)
7. Simonyan, K., Parkhi, O., Vedaldi, A., Zisserman, A.: Fisher vector faces in the wild. In: BMVC, vol. 2, p. 4 (2013)
8. Holub, A., Welling, M., Perona, P.: Combining generative models and fisher kernels for object class recognition. In: ICCV, pp. 136–143 (2005)
9. Perronnin, F., Dance, C.: Fisher kernels on visual vocabularies for image categorization. In: IEEE Conference on Computer Vision and Pattern Recognition, pp. 1–8. IEEE (2007)
10. Sanchez, J., Perronnin, F.: High-dimensional signature compression for large-scale image classification. IEEE Conference on Computer Vision and Pattern Recognition (CVPR) **2011**, 1665–1672 (2011)
11. Csurka, G., Perronnin, F.: Fisher vectors: beyond bag-of-visual-words image representations. In: Richard, P., Braz, J. (eds.) Computer Vision, Imaging and Computer Graphics Theory and Applications. Communications in Computer and Information Science, vol. 229, pp. 28–42. Springer, Berlin (2011)
12. Sanchez, J., Perronnin, F., Mensink, T., Verbeek, J.: Image classification with the fisher vector: theory and practice. Technical Report RR-8209, INRIA (2013)
13. Chatfield, K., Lempitsky, V., Vedaldi, A., Zisserman, A.: The devil is in the details: an evaluation of recent feature encoding methods. In: Proceedings of BMVC, pp. 76.1–76.12. https://doi.org/10.5244/C.25.76 (2011)

14. Krizhevsky, A., Sutskever, I., Hinton, G.: Imagenet classification with deep convolutional neural networks. In: Advances in Neural Information Processing Systems, pp. 1097–1105 (2012)
15. Azim, T.: Fisher kernels match deep models. Electron. Lett. **53**(6), 397–399 (2017)
16. Perronnin, F., Larlus, D.: Fisher vectors meet neural networks: a hybrid classification architecture. In: Proceedings of the IEEE Conference on Computer Vision and Pattern Recognition, pp. 3743–3752 (2015)
17. Huang, P.S., Avron, H., Sainath, T., et al.: Kernel methods match deep neural networks on TIMIT. In: IEEE International Conference on Acoustics, Speech and Signal Processing (ICASSP), pp. 205–209. IEEE (2014)
18. Zeiler, M., Fergus, R.: Visualizing and understanding convolutional networks. In: European Conference on Computer Vision, pp. 818–833. Springer (2014)
19. Yu Zhang, J.W., Cai, J.: Compact representation for image classification: to choose or to compress? In: Proceedings of the IEEE Conference on Computer Vision and Pattern Recognition, pp. 907–914 (2014)
20. Chen, S., Liu, H., Zeng, X., et al.: Local patch vectors encoded by fisher vectors for image classification. Information **9**(2), 38 (2018)
21. Fred, A., De Marsico, M.: Pattern recognition applications and methods, pp. 80–98
22. Oliveira, F., Levkowitz, H.: From visual data exploration to visual data mining: a survey. IEEE Trans. Visual Comput. Graphics **9**(3), 378–394 (2003)
23. Ham, J., Lee, D., Mika, S., et al.: A kernel view of the dimensionality reduction of manifolds. In Proceedings of the Twenty-first International Conference on Machine Learning, p. 47. ACM (2004)
24. Maaten, L., Postma, E.: Dimensionality reduction: a comparative. J. Mach. Learn. Res., 66–71 (2009)
25. Vlachos, M., Domeniconi, C., Gunopulos, D., et al.: Non-linear dimensionality reduction techniques for classification and visualization. In: Proceedings of the Eighth ACM SIGKDD International Conference on Knowledge Discovery and Data Mining, pp. 645–651. ACM (2002)
26. Jolliffe, I.: Principal component analysis: Wiley online library. Google Scholar (2002)
27. Torgerson, W.: Multidimensional scaling: I: theory and method. Psychometrika, 401–419 (1952)
28. Ben-Bassat, M.: Pattern recognition and reduction of dimensionality. Handb. Stat. **1982**, 773–910 (1982)
29. Kira, K., Rendell, L.: A practical approach to feature selection. In: Machine Learning Proceedings 1992, pp. 249–256. Elsevier (1992)
30. Robnik, M., Kononenko, G.: Theoretical and empirical analysis of relieff and rrelieff. Mach. Learn., 23–69 (2003)
31. Koller, D., Sahami, M.: Toward Optimal Feature Selection. Technical report, Stanford InfoLab (1996)
32. Peng, H., Long, F., Ding, C.: Feature selection based on mutual information criteria of max-dependency, max-relevance, and min-redundancy. IEEE Trans. Pattern Anal. Mach. Intell. **27**(8), 1226–1238 (2005)
33. Hall, M., Smith, L.: Feature selection for machine learning: comparing a correlation-based filter approach to the wrapper. FLAIRS Conference **1999**, 235–239 (1999)
34. Kohavi, R., John, G.: Wrappers for feature subset selection. Artif. Intell., 273–324 (1997)
35. Ahmed, S., Azim, T.: Compression techniques for deep fisher vectors. In: ICPRAM, pp. 217–224 (2017)
36. Ahmed, S., Azim, T.: Condensing Deep Fisher Vectors: To Choose or to Compress? In: ICPRAM. Extended Papers, Series: LNCS, Subseries: Image Processing, Computer Vision, Pattern Recognition, and Graphics, pp. 80–98. Springer, Cham (2017)
37. Gulgezen, G., Cataltepe, Z., Yu, L.: Stable and accurate feature selection. In: Machine Learning and Knowledge Discovery in Databases, pp. 455–468. Springer, Berlin (2009)
38. Maaten, L.: Learning a parametric embedding by preserving local structure. In: Artificial Intelligence and Statistics, pp. 384–391 (2009)

Chapter 5
Open Source Knowledge Base for Machine Learning Practitioners

Abstract In this chapter, we provide references to some of the most useful resources that could provide practitioners a quick start for learning and implementing a variety of deep learning models, kernel functions, Fisher vector encodings and feature condensation techniques. Not only can the users benefit from the open source codes, a rich collection of benchmark data sets and tutorials can provide them all the details to get hands on experience of the techniques discussed in this book. We have shared comparative analysis of the resources in tabular form so that users could pick the tools keeping in view their programming expertise, software/hardware dependencies and productivity goals.

Keywords Benchmark data sets · Deep learning frameworks
Kernel learning libraries · Dimensionality reduction toolboxes
Fisher kernel toolbox · GPU/CPU support · Operating system compatibility
Open source software

5.1 Benchmark Data Sets

In order to get started with deep learning or kernel based learning, one requires *standard* data sets used by various researchers to benchmark the efficacy of their algorithms. Benchmark data sets are usually well understood by the user community and offer the advantage of performance comparison of new techniques with the state of the art to assess if any progress has been made. It is better to take a start by using small data sets that could easily be downloaded and does not take long to fit or train the models. Another important aspect of the data set is its free access. Most of the academics and research institutes keep their data collections open source to increase transparency, participation and invite innovation. We have maintained a list of some popular open source benchmark data sets in Tables 5.1, 5.2 and 5.3 used by the computer vision and machine learning community to check the excellence of their developed algorithms on various problems.

Table 5.1 The table below lists some notable object recognition and classification data sets extensively used to benchmark the performance of deep models and kernel methods

Data set name	Author(s)	Number of classes	Data set size and image specifications
MNIST [1]	Yann LeCun	10 classes	60,000 train and 10,000 test examples, Image Resolution: 28×28
USPS Data Set [2]	Yann LeCun	10 classes	7291 train and 2007 test instances, Image Resolution: 16×16
The Street View House Numbers (SVHN) Data Set [3]	Yuval Netzer and Tao Wang	10 classes	Total images: 30,420, Train images:73257, Test images: 26032; there are 531,131 additional images that are somewhat less difficult samples used as extra training data
COIL Data Set • COIL-20 [4] • COIL-100 [5]	Sameer Nene, Shree Nayar and Hiroshi Mur	• 20 classes • 100 classes	• Total images:1,440, 72 images per class. Image Resolution: 32×32 • Total images: 7200; 72 images per class; Image Resolution: 128×128
STL-10 Data Set [6]	Adam Coates, Honglak Lee and Andrew Ng	10 classes and 100000 unlabelled images	500 train and 800 test images per class; Image Resolution: 96×96
CIFAR Data Set [7] • CIFAR10 • CIFAR100	Alex Krizhevsky, Vinod Nair and Geoffrey Hinton	• 10 classes • 100 classes	• 50000 train images and 10000 test images • 500 train images and 100 test images Image Resolution: 32×32
Caltech Data Set • Caltech-101 [8] • Caltech256 [9]	• Fei-Fei, Rob Fergus and Pietro Perona • Gregory Griffin, Alex Holub and Pietro Perona	• 101 classes • 256 classes	• 40–800 images per category • 30607 total images
Caltech Silhouettes [10]	Benjamin Marlin	101 categories	Image resolution: 28×28 or 16×16 having 100 training examples for each of the 101 classes and 6–400 examples per class for validation and test sets
PASCAL VOC Data Set [11]	Mark Everingham	20 classes	Train/validation set has 11,530 images containing 27,450 ROI annotated objects and 6,929 segmentations
ImageNet [12]	Jia Deng, Wei Dong and Richard Socher	1000 classes	1.2 million train images, 50,000 validation images and 100,000 test images, Image resolution varies

Table 5.2 This table enlists some popular benchmark data sets of facial images used for the tasks of face detection, face recognition, facial expression recognition and 3-D face reconstruction

Data set name	Author(s)/University	Number of classes	Data set size and image specification
ORL Data Set [13]	Engineering Department,Cambridge University	40 classes	Total 400 gray–scale images with image resolution of 92×112
UMist Faces [14]	Graham and Daniel	20 categories	Total 564 gray-scale; Image Resolution: 220×220
AR Face Database [15]	Aleix Martinez and Robert Benavente	126 categories	Total 3,016 images, Image Resolution: 768×576
CVL Database [16]	Peter Peer	114 categories	Total 114 coloured images, Image Resolution: 640×480
Multi-Pie Database [17]	Ralph Gross and Iain Matthews	337 categories	Contains more than 750,000 coloured images with size of 3072×2048
Face-in-Action Data Set [18]	Rodney Goh and Lihao Liu etal	180 categories	This database consists of 20sec videos of 24-bit coloured images of face data with resolution of 640×480 pixels
Facial Recognition Technology (FERET) Data Set [19]	Phillips and Harry Wechsler	24 categories	Total 14,126 images with resolution 256×384
• The Yale Face Database [20] • The Extended Yale Face Database B [21]	Athin Georghiades and Peter Belhumeur	• 15 categories • 28 categories	• 165 gray-scale images of resolution. 32×32 • 16,128 gray scale images of dimensionality 32×32

Table 5.3 The table below enlists some natural processing data sets to practice on when using deep learning models and kernel methods. These data sets have excessively been used in academic papers and can give a practitioner a quick start to understand the task at hand. More NLP data sets could be found here: https://machinelearningmastery.com/datasets-natural-language-processing/

Name	Author(s)	Number of classes	Data set size and specifications
20NewsGroup [22]	Ken Lang	20 groups	Total 18,846 documents after pre-processing; Train Documents: 11,308, Test Documents: 7,538
Reuters-21578 Corpus [23]	David Lewis	90 classes	7769 train documents and 3019 test documents
Large Movie Review Data Set [24]	Andrew Mass	2 classes	50,000 movie reviews, Train data: 25,000, Test data: 25,000

Table 5.4 The table below enlists the SVM toolboxes offering the functionality of classification/regression on various platforms. The practitioners can make a choice by keeping in view their programming expertise, hardware/software platforms on board and productivity goals

List of SVM toolboxes	Author(s)	Supported interface(s)	Operating system(s)	GPU/CPU implement
LIBLINEAR [25]	Rong-En Fan and Kai-Wei Chang	MATLAB,Octave, Python, R, Ruby, Perl and Scilab	Windows/Linux	CPU only
SVM^{light} Toolbox [26]	Thorsten Joachims	C, MATLAB, Python, Java and Ruby	Windows/Linux	CPU only
BudgetedSVM [27]	Nemanja Djuric and Liang Lan	C++	Windows	CPU only
Proximal Support Vector Machine (PSVM) [28]	Olvi Mangasarian, Edward Wild	MATLAB	Windows/Linux	CPU only
Divide-and-Conquer Kernel SVM (DC-SVM) [29]	Cho-Jui Hsieh, Si Si, and Inderjit Dhillon	MATLAB, C	Windows	CPU only
Least-square SVM (LS-SVMlab) [30]	Kristiaan Pelckmans, Johan Suykens	MATLAB	Windows/Linux	CPU only
SVM and Kernel Methods Matlab Toolbox [31]	Alain Rakotomamonjy and Stephane Canu	MATLAB	Windows/Linux	CPU only
Statistical Pattern Recognition Toolbox for MATLAB [32]	Vojtech Franc and Vaclav Hlavac	MATLAB and C	Windows/Linux	CPU only
SpiderSVM [33]	Jason Weston, Andre Elisseeff and Gokhan BakI	MATLAB	Windows/Linux	CPU only
BSVM [34]	Chih-Wei Hsu and Chih-Jen Lin	C++	Windows	CPU only
mySVM [35]	Stefan Ruping	C++	Windows/Linux	CPU only
LASVM [36]	Leon Bottou and Antoine Bordes	C	Windows	CPU only
SVMsequel [37]	Hal Daume III	OCaml	Linux	CPU only
SVMtorch [38]	Ronan Collobert and Samy Bengio	C++	Linux/Solaris	CPU only
LIBSVM [39]	Chih-Chung Chang and Chih-Jen Lin	MATLAB/Octave, Java, Python, R, C#, Ruby, PERL, CUDA, Scilab	Windows/Linux	CPU and GPU
Thunder SVM [40]	Zeyi Wen, Jiashuai Shi and Bingsheng	C/C++, Python, R, MATLAB	Linux/Windows/MAC	CPU and GPU
cuSVM [41]	Austin Carpenter	C	Windows/Linux	CPU and GPU
GTSVM [42]	Andrew Cotter,Nathan Srebro and Joseph Keshet	C/C++, MATLAB	Windows/Linux	GPU
Parallel GPDT [43]	Thomas Serafini,Luca Zanni and Gaetano Zanghirati	C++	Windows/Linux	GPU
GPUMlib [44]	Noel Lopes and Bernardete Ribeiro	C++	Windows	GPU
Rgtsvm [45]	Zhong Wang, Tinyi Chu and Lauren Choate	R	Linux/MAC OS	GPU

5.2 Standard Toolboxes and Frameworks: A Comparative Review

There is a wide range of open source machine learning libraries, toolboxes and frameworks that allow the machine learning engineers to build, extend and customise intelligent machine learning systems. We here provide a curated list of open source machine learning frameworks and toolboxes for implementing kernel learning, deep neural models and dimensionality reduction techniques. The use of open source libraries and software provides the practitioners the following benefits:

- No purchase cost and licensing restrictions.
- Well maintained developer and research community support.
- Standard documentation explaining the motivation behind development, providing installation guide and usage of code.
- Saves time to write the code from scratch.
- Makes performance comparison of the developed algorithm easier as the underlying framework of new and old softwares is the same.
- Provides flexibility to the programmers to code in the language of their own choice.
- Facilitate development of portable solutions on different computing platforms and devices.
- Enhances programmer's focus on solving the task at hand rather then getting stuck into the coding details.

Novice researchers or industrial practitioners having minimum programming expertise can benefit greatly from these listings by applying the models/algorithms to different real-life problems. We have compared some important features of each toolbox/framework in Tables 5.4, 5.5, 5.6 and 5.7 to make it easier for the readers to make a selection. If the data set in use is massive, it is advised to use a GPU based version of selected framework/toolbox that could run the software in parallel on multiple GPU cores. The software dependencies of each toolbox could be analysed by visiting the cited references of these listings.

It is also important to understand the difference between libraries and frameworks when selecting the tool of your choice. A library usually focuses on a single piece of functionality that the user can invoke through an application programming interface (API). The library executes a single or related set of functions and returns the control to the programmer. In contrast, the framework can have multiple libraries solving more than one functionality and the key attribute making it unique from the libraries is the presence of *inversion of control*. Inversion of control is a design principle where the custom written portion of user program receives flow of control from a generic framework that it respects. The abstractions provided by the framework need to be respected and the framework dictates which routine to run next rather than the user.

Table 5.5 List of well received toolboxes and frameworks for Fisher kernel, multiple kernel and other popular pattern recognition and machine learning techniques

Name of ToolBox	Author(s)	Operating system(s)	Programming language(s)
Vlfeat [46]	Andrea Vedaldi and Brian Fulkerson	MAC OS, Linux and Windows	MATLAB/Octave and C
Yael [47]	Herv Jegou and Matthijs Douze	MAC OS, Linux/Windows	MATLAB, Python and C
Fisher Kernel Learning Toolbox [48]	Laurens van der Maaten	MAC, Linux/Windows	MATLAB
Kernel Smoothing Toolbox [49]	Jan Kolacek and Jiri Zelinka	MAC, Linux/Windows	MATLAB
SHOGUN Toolbox [50]	Soren Sonnenburg, Gunnar Ratsch and Sebastian Henschel	GNU/Linux, Mac OSX, FreeBSD, and Windows	C++, MATLAB, R, Octave, Python, Java, Scala, Ruby, C#, R, Lua
OpenKernelLibrary [51]	Cyril Allauzen and Mehryar Mohri	Linux and MacOS X using g++	C++
DOGMA (online alg:UFO) [52]	Francesco Orabona	Linux/Mac OSX	MATLAB
MKL-SMO [53]	Vishwanathan and Manik Varma	Windows/Linux	C++
Multiple Kernel Learning Algorithms [54]	Mehmet Gonenn and Ethem Alpaydn	Windows/Linux	MATLAB
LIBLINEAR MKL [55]	Ming-Hen Tsai	Windows/ Linux	MATLAB/C++
Generalized Multiple Kernel Learning [56]	Manik Varma	Windows/Linux	MATLAB
Localized Algorithms for Multiple Kernel Learning [57]	Mehmet Gonen and Ethem Alpaydn	Windows/Linux	MATLAB
Deep multiple Kernel learning [58]	Eric Strobl and Shyam Visweswaran	Windows/Linux	MATLAB
Kernel Methods for Pattern Analysis MATLAB Toolbox [59]	John Shawe-Taylor	Windows/Linux	MATLAB
Pattern Recognition and Machine Learning Toolbox [60]	Mo Chen	Windows/Linux	MATLAB

Table 5.6 This table enlists various popular deep learning frameworks and libraries. The use of libraries and frameworks makes it easier for the practitioners to define any deep model or build wrappers around an architecture of their own choice. The libraries cited here offer limited functionality for a few deep models, whereas the frameworks are more extensive offering a variety of different deep neural models and learning techniques supported and optimised for various platforms and GPUs

Deep learning toolboxes and frameworks	Author(s)/Organisation	Operating System(s)	Programming language(s)
Libraries			
Deep Boltzmann Machine [61]	Ruslan Salakhutdinov	Windows,Linux and MAC OS	MATLAB
Restricted Boltzmann Machine	• Rasmusbergpalm [62] • Ruslan [63]	Windows, Linux and MAC OS	MATLAB
Deep Belief Networks [64]	William Gallamine	Windows, Linux and MAC OS	MATLAB
MATLAB Neural Network Toolbox [65]	Howard Demuth, Mark Beale	Windows, Linux and MACOS	MATLAB
DeepNetToolbox [66]	Nitish Srivastava	Windows, Linux and MAC OS	Python
CUDA-Convnet [67]	Alex Krizhevsky	Windows, MAC OS, Linux	C++/CUDA
FrameWorks			
TensorFlow [68]	Google Brain	Windows, MAC OS, UBUNTU	Java, Python, GO and C++
Torch [69]	Ronan Collobert and Kavukcuoglu	Linux, Android, MAC OS, iOS	C and LuaJIT
Microsoft Cognitive Toolkit (CNTK) [70]	Frank Seide	Windows and Linux	C++
Theano [71]	James Bergstra and Frederic Bastien	Ubuntu, MAC OS, Windows & CentOS 6.I	Python
Caffe [72]	Yangqing Jia and Evan Shelhame	Linux, MAC OS, Windows	C++, Python and MATLAB
Keras [73]	Franois Chollet	Linux, MAC OS, Windows, Android	Python
MXNet [74]	Tianqi Chen and Mu Li	Windows, MAC OS, Linux	C++, Python, R. Julia, Java Script, Scala, GO, Perl
Deeplearning4j [75]	Adam Gibson and Chris Nicholson	Linux, MAC OS, Windows, Android	Java, Scala Clojure and Python
Chainer [76]	Seiya Tokui (Preferred Networks)	Linux, OS, Windows	Python
Neon [77]	Nervana Systems	Linux and MAC OS	Python
LightNet [78]	Chengxi Ye, Chen Zhao, Yezhou Yang and Cornelia Fermlle	MAC OS, Linux and Windows	MATLAB
Apache Singa [79]	Beng Chin and Kian-Lee	Linux, MAC OS, Windows	Python, C++, Java

Table 5.7 List of famous feature compression and feature selection libraries. Each tool box offers a variety of different standard procedures for defining data, extracting features, changing feature representation spaces, density estimation, classifiers and evaluation metrics

Name of feature compression/Selection tool box	Author(s)	Feature selection/Compression techniques	Programming language(s)
YAN-PRTools [80]	Ke Yan	Selection & Compression	MATLAB
PRTools: A MATLAB Toolbox for Pattern Recognition [81]	Duin, Juszczak and Paclik.	Selection	MATLAB
Feature Selection Toolbox 3 (FST3) [82]	Petr Somol and Pavel Vacha	Selection	C++
Maximum-likelihood Feature Selection (MLFS) [83])	Sugiyama-Sato, Honda Lab @ University of Tokyo	Selection	MATLAB
Least-Squares Feature Selection (LSFS) [84]	Takafumi Kanamori and Masashi Sugiyama	Selection	MATLAB
L1-LSMI-Based Feature Selection (L1-LSMI) [85]	Wittawat Jitkrittum and Masashi Sugiyama	Selection	MATLAB
Feature Selection Library [86]	Giorgio Roffo	Selection	MATLAB
Dimensionality Reduction toolbox [87]	Laurens van der Maaten	Compression	MATLAB
Autoencoder [88]	Ruslan Salakhutdinov & Geoff Hinton	Compression	MATLAB
Parametric t-SNE [89]	Laurens van der Maaten	Compression	MATLAB
Neighborhood Preserving Embedding (NPE)[90]	Xiaofei He and Deng Cai	Compression	MATLAB
Locality Sensitive Discriminant Analysis (LSDA) [91]	Deng Cai and Xiaofei He.	Compression	MATLAB
Semi-Supervised Discriminant Analysis (SDA) [92]	Deng Cai, Xiaofei He and Jiawei Han	Compression	MATLAB
Maximum Margin Projection (MMP) [93]	Xiaofei He, Deng Cai and Jiawei Han	Compression	MATLAB
Least-Squares Dimensionality Reduction (LSDR) [94]	Taiji Suzuki and Masashi Sugiyama	Compression	MATLAB
Semi-Supervised LFDA (SELF) [95]	Masashi Sugiyama and Tsuyoshi Ide	Compression	MATLAB
Local Fisher Discriminant Analysis (LFDA) [96]	Masashi Sugiyama	Compression	MATLAB
Learning to Hash (Compression Methods Based on Hashing) [97]	Wu-Jun LI	Compression	MATLAB
Searching with Quantization Package (Product Quantization) [98]	Herve Jegou—Inria	Compression	MATLAB

References

1. LeCun, Y.: The MNIST database of handwritten digits (1998). http://yann.lecun.com/exdb/mnist/
2. Lecun, Y.: USPS dataset. http://www.cad.zju.edu.cn/home/dengcai/Data/MLData.html
3. Netzer, Y., Wang, T., Coates, A., et al.: Reading digits in natural images with unsupervised feature learning. In: NIPS Workshop on Deep Learning and Unsupervised Feature Learning, vol. 2011 (2011). http://ufldl.stanford.edu/housenumbers/
4. Nene, S., Nayar, S., Murase, H.: Columbia object image library (COIL-20) (1996). http://www.cs.columbia.edu/CAVE/software/softlib/coil-20.php
5. Nene, S., Nayar, S., Murase, H.: Columbia object image library (COIL-100). http://www.cs.columbia.edu/CAVE/software/softlib/coil-100.php
6. Coates, A., Lee, H., Ng, A.: An analysis of single-layer networks in unsupervised feature learning. In: Proceedings of the Fourteenth International Conference on Artificial Intelligence and Statistics, pp. 215–223 (2011). https://cs.stanford.edu/~acoates/stl10/
7. Krizhevsky, A., Nair, V., Hinton, G.: The CIFAR dataset (2014). https://www.cs.toronto.edu/~kriz/cifar.html
8. Fei-Fei, L., Fergus, R., Perona, P.: Learning generative visual models from few training examples: an incremental bayesian approach tested on 101 object categories. Comput. Vision Image Underst., 59–70 (2007). http://www.vision.caltech.edu/Image_Datasets/Caltech101/
9. Griffin, G., Holub, A., Perona, P.: Caltech-256 object category dataset (2007). http://www.vision.caltech.edu/Image_Datasets/Caltech256/
10. Marlin, B., Swersky, K., et al.: Inductive principles for restricted boltzmann machine learning. In: Proceedings of the Thirteenth International Conference on Artificial Intelligence and Statistics. pp. 509–516 (2010). https://people.cs.umass.edu/~marlin/data.shtml
11. Everingham, M., Gool, L., Williams, C., et al.: The pascal visual object classes (VOC) challenge. Int. J. Comput. Vision **88**, 303–338 (2010). http://host.robots.ox.ac.uk/pascal/VOC/
12. Deng, J., Dong, W., Socher, R., et al.: Imagenet: a large-scale hierarchical image database. In: IEEE Conference on Computer Vision and Pattern Recognition (CVPR), pp. 248–255. IEEE (2009). http://www.image-net.org/
13. Computer Laboratory Cambridge University: The ORL database of faces. http://www.cl.cam.ac.uk/research/dtg/attarchive/facedatabase.html
14. Graham, D., Allinson, N., et al.: Characterising virtual eigensignatures for general purpose face recognition. In: Face Recognition, pp. 446–456. Springer (1998). https://cs.nyu.edu/~roweis/data.html
15. Martinez, A., Benavente, R.: The AR face database, 1998. Comput. Vision Cent. Technical Report **3** (2007). http://www2.ece.ohio-state.edu/~aleix/ARdatabase
16. Peer, P.: CVL face database. Computer Vision Lab, Faculty of Computer and Information Science, University of Ljubljana, Slovenia (2005). http://www.lrv.fri.uni-lj.si/facedb.html
17. Gross, R., Matthews, I., Cohn, J., et al.: The CMU multi-pose, illumination, and expression (Multi-PIE) face database. CMU Robotics Institute. TR-07-08, Technical Report (2007). http://www.cs.cmu.edu/afs/cs/project/PIE/MultiPie/Multi-Pie/Home.html
18. Goh, R., Liu, L., Liu, X.: The CMU face in action (FIA) database. In: International Workshop on Analysis and Modeling of Faces and Gestures, pp. 255–263. Springer (2005). https://www.flintbox.com/public/project/5486/
19. Phillips, P., Wechsler, H., Huang, J., Rauss, P.: The FERET database and evaluation procedure for face-recognition algorithms. Image Vision Comput. **16**, 295–306 (1998). https://www.nist.gov/itl/iad/image-group/color-feret-database
20. Georghiades, A., Belhumeur, P., Kriegman's, D.: The yale face database. http://vision.ucsd.edu/datasets/yale_face_dataset_original/yalefaces.zip
21. Georghiades, A., Belhumeur, P., Kriegman's, D.: From few to many: illumination cone models for face recognition under variable lighting and pose. IEEE Trans. Pattern Anal. Mach Intell. **23**, 643–660 (2001). http://vision.ucsd.edu/~iskwak/ExtYaleDatabase/ExtYaleB.html

22. Lang, K.: The 20 newsgroups dataset. http://qwone.com/~jason/20Newsgroups/
23. Lewis, D.: Reuters-21578 dataset. http://www.daviddlewis.com/resources/testcollections/
    reuters21578/
24. Maas, A., Daly, R., Pham, P., et al.: Learning word vectors for sentiment analysis. In: Proceedings of the 49th Annual Meeting of The Association for Computational Linguistics: Human Language Technologies, Portland, Oregon, USA, Association for Computational Linguistics, June 2011, pp. 142–150 (2011). http://www.aclweb.org/anthology/P11-1015
25. Fan, R., Chang, K., Hsieh, C., Wang, X., Lin, C.: LIBLINEAR: a library for large linear classification. J. Mach. Learn. Res. **9**, 1871–1874 (2008)
26. Joachims, T.: SVMlight: support vector machine. **19**(4) (1999). http://svmlight.joachims.org/
27. Djuric, N., Lan, L., Vucetic, S.: BudgetedSVM: a toolbox for scalable SVM approximations. J. Mach. Learn. Res., 3813–3817 (2013). https://sourceforge.net/p/budgetedsvm/code/ci/master/tree/matlab/
28. Mangasarian, O., Wild, E.: Proximal support vector machine classifiers. In: Proceedings KDD-2001: Knowledge Discovery and Data Mining, pp. 77–86 (2001). http://research.cs.wisc.edu/dmi/svm/psvm/
29. Hsieh, C., Si, S., Dhillon, I.: A Divide-and-conquer solver for kernel support vector machines. In: International Conference on Machine Learning (2014). http://www.cs.utexas.edu/~cjhsieh/dcsvm/
30. Suykens, J., Pelckmans, K.: Least squares support vector machines. Neural Process. Lett., 293–300 (1999). https://www.esat.kuleuven.be/sista/lssvmlab/
31. Rakotomamonjy, A., Canu, S.: SVM and kernel methods MATLAB toolbox (2008). http://asi.insa-rouen.fr/enseignants/~arakoto/toolbox/
32. Franc, V., Hlavac, V.: Statistical pattern recognition toolbox for MATLAB. Prague, Czech: Center for Machine Perception, Czech Technical University (2004). https://cmp.felk.cvut.cz/cmp/software/stprtool/
33. Weston, J., Elisseeff, A., Bak, G.: Spider SVM toolbox (2006). http://people.kyb.tuebingen.mpg.de/spider/
34. Hsu, C.W., Lin, C.J.: BSVM-2.06 (2009). https://www.csie.ntu.edu.tw/~cjlin/bsvm/
35. Ruping, S.: Mysvm–a support vector machine (2004). http://www-ai.cs.uni-dortmund.de/SOFTWARE/MYSVM/index.html
36. Bottou, L., Bordes, A., Ertekin, S.: Lasvm (2009). http://leon.bottou.org/projects/lasvm#introduction
37. III, H.D.: SVMseq documentation. http://legacydirs.umiacs.umd.edu/~hal/SVMsequel/
38. Collobert, R., Bengio, S.: SVMTorch: support vector machines for large-scale regression problems. J. Mach. Learn. Res. (2001). http://bengio.abracadoudou.com/SVMTorch.html
39. Chang, C., Lin, C.: LIBSVM: a library for support vector machines. ACM Trans. Intell. Syst. Technol., 27:1–27:27 (2011). http://www.csie.ntu.edu.tw/~cjlin/libsvm
40. Wen, Z., Shi, J., He, B., et al.: ThunderSVM: a fast SVM library on GPUs and CPUs. https://github.com/zeyiwen/thundersvm
41. Carpenter, A.: CUSVM: a CUDA implementation of support vector classification and regression, pp. 1–9 (2009). http://patternsonascreen.net/cuSVM.html
42. Cotter, A., Srebro, N., Keshet, J.: A GPU-tailored apppproach for training kernelized SVM. In: Proceedings of the 17th ACM SIGKDD international conference on knowledge discovery and data mining, pp. 805–813 (2011). http://ttic.uchicago.edu/~cotter/projects/gtsvm/
43. Serafini, T., Zanni, L., Zanghirati, G.: Parallel GPDT: a parallel gradient projection-based decomposition technique for support vector machines (2004). http://dm.unife.it/gpdt/
44. Lopes, N., Ribeiro, B.: GPUMLib: a new library to combine machine learning algorithms with graphics processing units. In: 2010 10th International Conference on Hybrid Intelligent Systems (HIS), pp. 229–232 (2010). https://sourceforge.net/projects/gpumlib/?source=typ_redirect
45. Wang, Z., Chu, T., Choate, L., et al.: Rgtsvm: support vector machines on a GPUin R. ArXiv Preprint ArXiv:1706.05544 (2017). https://github.com/Danko-Lab/Rgtsvm

46. Vedaldi, A., Fulkerson, B.: VLFeat: an open and portable library of computer vision algorithms. In: Proceedings of the 18th ACM International Conference on Multimedia, pp. 1469–1472 (2010). http://www.vlfeat.org/install$-$matlab.html

47. Jegou, H., Douze, M.: The yael library. In: Proceedings of the 22nd ACM International Conference on Multimedia, pp. 687–690 (2014). https://gforge.inria.fr/projects/yael/

48. Maaten, L.: Fisher kernel learning. https://lvdmaaten.github.io/fisher/Fisher_Kernel_Learning.html

49. Kolacek, J., Zelinka, J.: Kernel smoothing in MATLAB: theory and practice of kernel smoothing (2012). http://www.math.muni.cz/english/science-and-research/developed-software/232-matlab-toolbox.html

50. Sonnenburg, S., Ratsch, G., Henschel, S.: J. Mach. Learn. Res., y.n.: The SHOGUN Machine Learning Toolbox

51. Allauzen, C., Mohri, M., Rostamizadeh, A.: Openkernel library (2007). http://www.openkernel.org/twiki/bin/view/Kernel/WebHome

52. Orabona, F.: DOGMA: A MATLAB toolbox for online learning (2009). http://dogma.sourceforge.net

53. Sun, Z., Ampornpunt, N., Varma, M., Vishwanathan, S.: Multiple kernel learning and the SMO algorithm. In: Advances in Neural Information Processing Systems (2010). http://manikvarma.org/code/SMO-MKL/download.html

54. Gonen, M., Alpaydin, E.: Multiple kernel learning algorithms. J. Mach. Learn. Res. (2011). https://users.ics.aalto.fi/gonen/jmlr11.php

55. Tsai, M.H.: LIBLINEAR MKL: a fast multiple kernel learning L1/L2-loss SVM solver in MATLAB. https://www.csie.ntu.edu.tw/~b95028/software/liblinear-mkl/

56. Varma, M., Babu, R.: More generality in efficient multiple kernel learning. In: Proceedings of the 26th Annual International Conference on Machine Learning, pp. 1065–1072 (2009). http://manikvarma.org/code/GMKL/download.html

57. Gonen, M., Alpaydn, E.: Localized algorithms for multiple kernel learning. Pattern Recognit. (2013). https://users.ics.aalto.fi/gonen/icpr10.php

58. Strobl, E., Visweswaran, S.: Deep multiple kernel learning. In: 2013 12th International Conference on Machine Learning and Applications (ICMLA) (2013). https://github.com/ericstrobl/deepMKL

59. Shawe-Taylor, J.: Kernel methods for pattern analysis (2004). https://www.kernel-methods.net/matlab_tools

60. Chen, M.: Pattern recognition and machine learning toolbox. MATLAB Central File Exchange (2016). https://www.mathworks.com/matlabcentral/fileexchange/55826-pattern-recognition-and-machine-learning-toolbox

61. Salakhutdinov, R., Hinton, G.: Deep Boltzmann machines. In: Proceedings of the International Conference on Artificial Intelligence and Statistics, vol. 5, pp. 448–455 (2009). http://www.cs.toronto.edu/~rsalakhu/DBM.html

62. Rasmusbergpalm: Restricted Boltzmann Machine. https://code.google.com/archive/p/matrbm/

63. Salakhutdinov, R., Hinton, G.: Restricted Boltzmann machines for collaborative filtering. In: Proceedings of the 24th International Conference on Machine Learning, pp. 791–798 (2007). http://www.cs.toronto.edu/~rsalakhu/rbm_ais.html/

64. Gallamine, W.: Deep belief network. https://github.com/gallamine/DBN

65. Demuth, H., Beale, M.: Neural Network Toolbox for Use with Matlab–User's Guide verion 3.0. (1993). https://www.mathworks.com/help/nnet/getting-started-with-neural-network-toolbox.html

66. Srivastava, N.: DeepNet: a library of deep learning algorithms. http://www.cs.toronto.edu/~nitish/deepnet

67. Krizhevsky, A.: Cuda-convnet: High-performance C++/Cuda implementation of convolutional neural networks (2012). https://code.google.com/archive/p/cuda-convnet2/

68. Abadi, M., Barham, P., et al.: TensorFlow: a system for large-scale machine learning. In: OSDI, vol. 16, pp. 265–283 (2016). https://www.tensorflow.org/

69. Collobert, R., Kavukcuoglu, K., Farabet, C.: Torch. In: Workshop on Machine Learning Open Source Software, NIPS, vol. 113 (2008). http://torch.ch/
70. Seide, F.: Keynote: the computer science behind the Microsoft cognitive toolkit: an open source large-scale deep learning toolkit for windows and linux. In: IEEE/ACM International Symposium on Code Generation and Optimization (CGO), pp. xi–xi (2017). https://www.microsoft.com/en-us/cognitive-toolkit/
71. Bergstra, J., Bastien, F., et al.: Theano: deep learning on GPUS with python. In: NIPS 2011, Big Learning Workshop, Granada, Spain, vol. 3, pp. 1–48 (2011). http://deeplearning.net/software/theano/
72. Jia, Y., Shelhamer, E., et al.: Caffe: convolutional architecture for fast feature embedding. In: Proceedings of the 22nd ACM International Conference on Multimedia, pp. 675–678 (2014) http://caffe.berkeleyvision.org/
73. Chollet, F.: Keras (2015). https://keras.io/
74. Chen, T., Li, M., Li, Y., et al.: Mxnet: A flexible and efficient machine learning library for heterogeneous distributed systems. ArXiv Preprint ArXiv:1512.01274 (2015). https://mxnet.apache.org/
75. Gibson, A., Nicholson, C., Patterson, J.: Deeplearning4j: open-source distributed deep learning for the JVM. Apache Softw. Found. License 2 (2016). https://deeplearning4j.org/
76. Tokui, S., Oono, K., Hido, S.: Chainer: a next-generation open source framework for deep learning. In: Proceedings of Workshop on Machine Learning Systems in The Twenty-Ninth Annual Conference on Neural Information Processing Systems (NIPS) (2015). https://chainer.org,
77. Neon, N.: Nervana systems. https://neon.nervanasys.com/index.html/
78. Ye, C., Zhao, C., Yang, Y., Fermlle, C.: Lightnet: a versatile, standalone matlab-based environment for deep learning. In: Proceedings of the 2016 ACM on Multimedia Conference, pp. 1156–1159 (2016). https://github.com/yechengxi/LightNet
79. Chin, B., Lee, K., Wang, S., et al.: SINGA: a distributed deep learning platform. In: Proceedings of the 23rd ACM International Conference on Multimedia. pp. 685–688 (2015). https://singa.incubator.apache.org/en/index.html
80. Yan, K.: Feature selection toolbox. https://www.mathworks.com/matlabcentral/fileexchange/56723-yan-prtools
81. Duin, R.P.W.: Prtools Version 3.0: a matlab toolbox for pattern recognition. In: Proceedings of the SPIE (2000). http://prtools.org/software/
82. Somol, P., Vacha, P., Mikes, S., et al.: Introduction to feature selection toolbox 3–the C++ library for subset search, data modeling and classification. Research Report for Institute of Information Theory and Automation, Academy of Sciences of the Czech Republic (2010). http://fst.utia.cz/?fst3
83. Lal, S.S.H.: Maximum likelihood feature selection (MLFS), University of Tokyo. http://www.ms.k.u-tokyo.ac.jp/software.html#MLFS
84. Karamori, T., Sugiyama, M.: A least-squares approach to direct importance estimation. J. Mach. Learn. Res. 10, 1391–1445 (2009). http://www.ms.k.u-tokyo.ac.jp/software.html#LSFS
85. Jitkrittum, W., Sugiyama, M.: Feature selection Via L1-penalized squared-loss mutual information. IEICE Trans. Inf. Syst. 96, 1513–1524 (2013). http://wittawat.com/pages/l1lsmi.html
86. Roffo, G.: Feature selection library (MATLAB Toolbox). ArXiv Preprint ArXiv:1607.01327 (2016). https://www.mathworks.com/matlabcentral/fileexchange/56937-feature-selection-library
87. Maaten, L.: Matlab toolbox for dimensionality reduction (2007). https://lvdmaaten.github.io/drtoolbox
88. Salakhutdinov, R., Hinton, G.: Reducing the dimensionality of data with neural networks. Science 313, 504–507 (2006). http://www.cs.toronto.edu/~hinton/MatlabForSciencePaper.html
89. Maaten, L.: Learning a parametric embedding by preserving local structure. In: Artificial Intelligence and Statistics, pp. 384–391 (2009). https://lvdmaaten.github.io/tsne/
90. He, X., Cai, D., et al.: Neighborhood preserving embedding. In: Tenth IEEE International Conference on Computer Vision, vol. 2, pp. 1208–1213 (2005). http://www.cad.zju.edu.cn/home/dengcai/Data/DimensionReduction.html

91. Cai, D., He, X., Zhou, K., Han, J., Bao, H.: Locality sensitive discriminant analysis. In: International Joint Conference on Artificial Intelligence (2007). http://www.cad.zju.edu.cn/home/dengcai/Data/DimensionReduction.html

92. Cai, D., He, X., Han, J.: Semi-supervised discriminant analysis. In: Proceedings of International Conference on Computer Vision (2007). http://www.cad.zju.edu.cn/home/dengcai/Data/DimensionReduction.html

93. He, X., Cai, D., Han, J.: Learning a maximum margin subspace for image retrieval. IEEE Trans. Knowl. Data Eng. **20** (2008). http://www.cad.zju.edu.cn/home/dengcai/Data/DimensionReduction.html

94. Suzuki, T., Sugiyama, M.: Sufficient dimension reduction via squared-loss mutual information estimation, pp. 804–811 (2010). http://www.ms.k.u-tokyo.ac.jp/software.html#LSDR

95. Sugiyama, M., Ide, T., et al.: Semi-supervised local fisher discriminant analysis for dimensionality reduction. Mach. Learn. **78**, 35 (2010). http://www.ms.k.u-tokyo.ac.jp/software.html#SELF

96. Sugiyama, M.: Local fisher discriminant analysis for supervised dimensionality reduction. In: Proceedings of the 23rd International Conference on Machine Learning, pp. 905–912. ACM (2006). http://www.ms.k.u-tokyo.ac.jp/software.html#LFDA

97. LI, W.: Learning to hashing. https://cs.nju.edu.cn/lwj/L2H.html

98. Jegou, H., Douze, M., Schmid, C.: Product quantization for nearest neighbor search. IEEE Trans. Pattern Anal. Mach. Intell. **33**(1), 117–128 (2011). http://people.rennes.inria.fr/Herve.Jegou/projects/ann.html